Vascular Endothelium

Physiological Basis of Clinical Problems II

NATO ASI Series

Advanced Science Institutes Series

A series presenting the results of activities sponsored by the NATO Science Committee, which aims at the dissemination of advanced scientific and technological knowledge, with a view to strengthening links between scientific communities.

The series is published by an international board of publishers in conjunction with the NATO Scientific Affairs Division

A	**Life Sciences**	Plenum Publishing Corporation
B	**Physics**	New York and London
C	**Mathematical and Physical Sciences**	Kluwer Academic Publishers
D	**Behavioral and Social Sciences**	Dordrecht, Boston, and London
E	**Applied Sciences**	
F	**Computer and Systems Sciences**	Springer-Verlag
G	**Ecological Sciences**	Berlin, Heidelberg, New York, London,
H	**Cell Biology**	Paris, Tokyo, Hong Kong, and Barcelona
I	**Global Environmental Change**	

Recent Volumes in this Series

Volume 254 —Slow Potential Changes in the Human Brain
edited by W. C. McCallum and S. H. Curry

Volume 255 —Quantitative Assessment in Epilepsy Care
edited by Harry Meinardi, Joyce A. Cramer, Gus A. Baker, and
Antonio Martins da Silva

Volume 256 —Advances in the Biomechanics of the Hand and Wrist
edited by F. Schuind, K. N. An, W. P. Cooney III, and
M. Garcia-Elias

Volume 257 —Vascular Endothelium: Physiological Basis of Clinical Problems II
edited by John D. Catravas, Allan D. Callow, and Una S. Ryan

Volume 258 —A Multidisciplinary Approach to Myelin Diseases II
edited by S. Salvati

Volume 259 —Experimental and Theoretical Advances in Biological Pattern
Formation
edited by Hans G. Othmer, Philip K. Maini, and James D. Murray

Volume 260 —Lyme Borreliosis
edited by John S. Axford and David H. E. Rees

Series A: Life Sciences

Vascular Endothelium

Physiological Basis of Clinical Problems II

Edited by

John D. Catravas

Medical College of Georgia
Augusta, Georgia

Allan D. Callow

Washington University Medical School
St. Louis, Missouri

and

Una S. Ryan

T Cell Sciences, Inc.
Cambridge, Massachusetts

Springer Science+Business Media, LLC

Proceedings of a NATO Advanced Study Institute on
Vascular Endothelium: Physiological Basis of Clinical Problems II,
held June 20–30, 1992,
in Rhodes, Greece

Library of Congress Cataloging-in-Publication Data

Vascular endothelium : physiological basis of clinical problems II /
   [edited by] John D. Catravas, Allan D. Callow, and Una S. Ryan.
       p.   cm. -- (NATO ASI series. Series A, Life sciences ; v.
   257)
     "Proceedings of a NATO Advanced Study Institute on Vascular
   Endothelium: Physiological Basis of Clinical Problems II, held June
   20-30, 1992, in Rhodes, Greece"--T.p. verso.
     "Published in cooperation with NATO Scientific Affairs Division."
   Includes bibliographical references and index.
     ISBN 978-0-306-44633-7   ISBN 978-1-4615-2437-3 (eBook)
     DOI 10.1007/978-1-4615-2437-3
     1. Vascular endothelium--Pathophysiology--Congresses.
   I. Catravas, John D. II. Callow, Allan D. III. Ryan, Una S., 1941-
   . IV. North Atlantic Treaty Organization. Scientific Affairs
   Division. V. NATO Advanced Study Institute on Vascular Endothelium:
   Physiological Basis of Clinical Problems II (1992 : Rhodes, Greece)
   VI. Series.
     [DNLM: 1. Endothelium, Vascular--physiopathology--congresses.   QS
   532.5.E7 V3318 1992]
   RC691.4.V38   1993
   616.1'307--dc20
   DNLM/DLC   93-27610
   for Library of Congress   CIP

ISBN 978-0-306-44633-7

PREFACE

This book is a compilation of the lectures and oral and poster communications presented at the Advanced Study Institute on *"Vascular Endothelium: Physiological Basis of Clinical Problems II,"* which took place between June 20 and 30, 1992 in Rhodes, Greece. This third in a series of ASIs on vascular endothelium continued on the theme of the first (1988) ASI on *"Receptors and Transduction Mechanisms"* and particularly expanded that of the 1990 conference on *"Physiological Basis of Clinical Problems."* We continued the successful practice of bringing together clinicians and scientists: this was reflected equally well in the composition of the organizing committee as in the background of the participants. Endothelial cell functions and dysfunctions present as many challenges to the investigator as they do to the curious clinical practitioner. As these problems are necessarily different, this unique ten-day co-habitation of these individuals continued to offer fresh outlooks to each, stimulated potential collaborative efforts and, most importantly, advanced --ever so slightly-- our knowledge of vascular biology. This year's conference was further enriched by the presence of several of our colleagues from Eastern Europe whom we are delighted to welcome as officially sponsored participants to this and future NATO-supported meetings.

It is never superfluous to remind readers and participants that those signing at the bottom of this page, while responsible for many of the ASI's and the book's deficiencies, are but three of the many contributors to the successes. We continue to be impressed by the high and sustained interest and interaction of the participants, reflected not only in the long formal --and after hours-- discussion periods, but also on their numerous and high quality presentations. Their enthusiasm and kind words of praise are the major forces sustaining the continuation of these conferences. Similarly, we are fortunate to depend upon individuals such as Norman Gillis, Alberto Mantovani and Magdi Yacoub for their critical contributions in putting together a stellar cast of lecturers and lectures that formed the nucleus of the ASI. We are equally indebted to the local organizing committee of Panaviotis Behrakis, Michael Maragoudakis and Lydia Argyropoulos for their tireless efforts before and during the meeting. Similarly, we wish to thank Mary Snead, James Parkerson, Andreas Papapetropoulos, Stelios Orfanos and Nandor Marczin for their considerable help in the preparation of the conference. This year we were fortunate to have Mary Ann Roupp responsible for both the processing of this book and the coordination of the ASI. We wish to add to the praise offered by so many participants for the quiet, efficient and friendly manner in which Ms. Roupp conducted the daily business of the conference, as well as for the considerable work that she undertook for about a year before and after the meeting. We are equally thankful for her exemplary editorial work, which is evident in this book.

John Catravas (Augusta)
Allan Callow (St. Louis)
Una Ryan (St. Louis)

CONTENTS

ABBREVIATIONS

β-gal	β-galactosidase
ACE	angiotensin converting enzyme
ALB	albumin
AMP	adenosine 5´ - monophosphate
ANKA	
BA	bovine aortic endothelial cells
BPAE	bovine pulmonary arterial endothelial cells
CAD	coronary artery disease
CAM	chorioallantoic membrane
CAPD	chronic ambulatory peritoneal dialysis
cGMP	cyclic guanosine 5´-monophosphate
CM	cerebral malaria
CMV	cytomegalovirus
cNOS	constitutive nitric oxide synthase
CRP	C-reactive protein
CSF	colony - stimulating factor
CTX	cytotoxicity
CYSNO	S-nitroso-L-cysteine
DG	diacylglycerol
EC	endothelial cells
ECE	endothelin converting enzyme
EDNO	endothelium-derived nitric oxide
EDRF	endothelium derived relaxing factor
ELAM	endothelial - leukocyte adhesion molecule
ELISA	enzyme-linked immuno-absorbance assay
ET	endothelin
FDD	flow-dependent dilator response
G	granulocyte
GAF	guanylyl cyclase activating factor
GC	guanylate cyclase
GM	granulocyte-macrophage
GMP	guanosine-5´-monophosphate
HAEC	human aortic endothelial cells
HB	hygromycin-B
HDL	high density lipoproteins
HUVEC	human umbilical vein endothelial cells
IBMX	isobutylmethyl xanthine
ICAM-1	intercellular adhesion molecule
IL	interleukin
IRF	impaired renal function

L-NAG	L-nitro-arginine
L-NMMA	L-mono-methyl-arginine
L-NNA	L-amino-arginine
LDL	low density lipoproteins
LNNA	L-N^G-nitro-arginine
LPS	lipopolysaccharide
LTBP	latent TGF-binding protein
LTR	long terminal repeat
M	monocyte
MB	methylene blue
MPANO	S-nitrose-3-mercaptoproprionic acid
MW	molecular weight
NANC	nonadrenergic-noncholinergic
NAR	nagase analbuminemic rats
NEM	N-ethyl-maleimide
NO	nitric oxide
ox-LDL	oxidized low-density lipoproteins
PAEC	pig aortic endothelial cells
PAF	platelet activating factor
PCR	polymerase chain reaction
PDGF	platelet-derived growth factor
PKC	protein kinase C
PMA	phorbol 12-myristate 13-acetate, TPA
PRBC	parasitized red blood cells
RBC	red blood cells
RPASM	rabbit pulmonary arterial smooth muscle
SAP	serum amyloid P component
SLe^x	sialyl-Lewis^x
SM	smooth muscle
SMC	smooth muscle cells
SNP	sodium nitroprusside
SOD	superoxide dismutase
TGFβ^1	transforming growth factorβ^1
TNF	tumor necrosis factor
TS	thrombospondin
UV	ultraviolet
VCAM-1	vascular cell adhesion molecule
VLA-4	very late antigen-4
WBC	white blood cells
XO	xanthine oxidase

I.   FREE RADICALS/ISCHEMIA REPERFUSION INJURY

BASAL RELEASE OF EDRF UNDER CONDITIONS OF OXIDANT STRESS

Nandor Marczin, Xilin Chen and John D. Catravas

Department of Pharmacology and Toxicology
Medical College of Georgia
Augusta, Georgia 30912, U.S.A.

INTRODUCTION

Tremendous progress has been made in our understanding of the nature as well as the mechanisms of generation and release of endothelium-derived relaxing factor (EDRF) since the milestone discovery of Furchgott and colleagues concerning the central role of endothelium in regulating vascular tone (*Furchgott and Zawadzki, 1980*). It is now evident that nitric oxide (NO) generated within the endothelial cells from the guanidino nitrogen atom of L-arginine via the constitutive activity of nitric oxide synthase (cNOS) conveys a variety of messages to maintain the fluidity of the blood, to prevent adhesion and aggregation of platelets and to determine the tone of the vessel wall by relaxing vascular smooth muscle (*Furchgott, 1988; Ignarro et al, 1988; Palmer et al, 1987; Bredt et al, 1991, Förstermann et al, 1991; Radomski et al, 1987*). Although the release of EDRF was initially described as a mode of action of pharmacological agents commonly acting on the surface of endothelial cells and increasing intracellular free calcium levels, agonist-independent formation, release and action of EDRF has recently also received considerable attention. Basal release of nitrogen oxides has been observed from cultured endothelial cells, and the ability of the endothelium to respond with increased EDRF output to mechanical stimuli such as pulsatility and increased blood flow has been demonstrated (*Schmidt et al., 1990; Myers et al., 1989; Rubanyi et al., 1986*). The introduction of specific inhibitors of endothelial NO synthesis proved to be an especially useful tool to delineate the *in vivo* activity and function of such NO regulation. Intravenous infusion of $L-N^G$-monomethyl-arginine (LNMMA) elicits a sustained elevation in arterial blood pressure in the guinea pig, rat and rabbit, and produces a decrease in vascular conductance in different vascular beds such as cerebral, coronary, pulmonary, mesenteric and renal microcirculation (*Aisaka et al., 1989; Gardiner et al., 1990; Rees et al., 1989; Tolins et al., 1990; Wiklund et al., 1990; Bassenge, 1991, Chu et al., 1991*). Likewise, intraarterial application of LNMMA in the forearm of normal subjects decreases local blood flow (*Vallance et al, 1989*). Similarly to LNMMA, hemoglobin, another inhibitor of NO, also increases blood pressure in healthy human subjects (*Savitsky et al., 1978.*) Treatment with NO inhibitors not only alters hemodynamics but also increases accumulation of platelets and neutrophils in the pulmonary circulation (*May et al., 1991*). These data clearly suggest an important endogenous regulatory role of the continuous tonic release of EDRF in the maintenance of systemic and regional blood pressure and flow, and antithrombogenicity of the vascular wall.

In light of these important functions of EDRF, it has been hypothesized that interference with NO action would result in an imbalance between the simultaneously present vasoconstrictors and dilators, and thrombogenic and antithrombogenic substances that might contribute to the development of vasospasm, hypertension and thrombosis. In fact, endothelial dysfunction is a prominent feature of atherosclerosis, hypertension, diabetes and ischemia-reperfusion, and the NO regulatory mechanism is impaired in all of these syndromes (for review, see *Marshall and Kontos, 1990*). One potentially common characteristic of these conditions is the increased formation of reduced oxygen species. Since some of these oxidants have the potential to interact with NO and eliminate its biological activity, as well as to injure endothelial and smooth muscle cells, oxidant-mediated inhibition of the synthesis, release and action of EDRF might be an important mechanism during vascular pathologic conditions. To better understand the nature of the interactions between reduced oxygen species and EDRF, we investigated changes in the biological activity of basally released EDRF under a variety of oxidant stress conditions.

AGONIST-INDEPENDENT, BASAL RELEASE OF EDRF IN CULTURE

The key discovery of Holtzman, Rapoport and Murad and Ignarro that vasodilation induced by EDRF can be correlated well with increases in cGMP levels in coronary arterial strips in the rat aorta and bovine intrapulmonary arteries offered a simple sensitive and specific biochemical assay to monitor the activity of EDRF in cell culture (*Holzmann, 1982; Rapoport and Murad, 1983; Ignarro et al., 1984*). On the basis of increased intracellular cGMP concentration in mixed cultures of endothelial cells and vascular smooth muscle cells, but not in either cells alone, Loeb and co-workers provided evidence for agonist-stimulated release of EDRF in cultured cells (*Loeb et al., 1985*). Shortly thereafter, Ganz and associates developed a long-term (2-3 days) and short-term (4 hours) co-culture model to study interactions between endothelial cells and smooth muscle cells. They reported the constitutive basal release of an endothelial humoral factor having properties similar to EDRF that increased basal cGMP levels in the smooth muscle cells (*Ganz et al., 1986*). We and others adopted these simple models and with the availability of specific inhibitors of endothelial NO synthesis proved that the factor responsible for endothelial-induced increases in co-culture cGMP levels completely depends on L-arginine-derived NO synthesis. In agreement with the above studies, we did not find any effect of L-arginine analogues on cGMP accumulation in single cultures of either endothelial cells, or smooth muscle cells (Figure 1A). In the majority of experiments with endothelial cells, this effect is due to the lack of activatable soluble guanylate cyclase and in smooth muscle cells this negative result is consistent with the lack of constitutive NO synthase. However, nitrovasodilators produce large increases in cGMP accumulation in smooth muscle cells, and the presence of endothelial cells increases basal cGMP levels in smooth muscle cells in both long-term (2-3 days) and in a short-term (15 min) co-culture models (Figure 1B; *Marczin et al., 1992a*). Importantly, these increases are completely prevented by LNMMA or L-N^G-nitro-arginine, LNNA (Figure 1B and Figure 2). Furthermore, the inhibitory action of LNMMA and LNNA can be reversed by L-arginine and L-arginine itself facilitates cGMP responses, especially in the long-term co-culture model or after depletion of endogenous L-arginine stores. This basal and agonist-stimulated activity was observed in a variety of macrovascular endothelial cells, although the magnitude of cGMP increases was variable among the cells from different species and vascular beds and was dependent upon the sGC activity of the smooth muscle cells used (Figure 2A and B).

SUPEROXIDE ANION-INDUCED INHIBITION OF EDRF ACTION

Superoxide anion is capable of inactivating nitric oxide and EDRF released after stimulation of endothelial cells with endothelium-dependent vasodilators (*Wei et al., 1985; Rubanyi and Vanhoutte, 1986; Gryglewski et al., 1986*). We have investigated interactions between superoxide anion and agonist-independent basal release of EDRF or exogenous nitrovasodilators under conditions of oxidant stress elicited by 1) methylene blue, 2) xanthine oxidase and 3) phorbol-myristate acetate (PMA)-activated neutrophils.

A

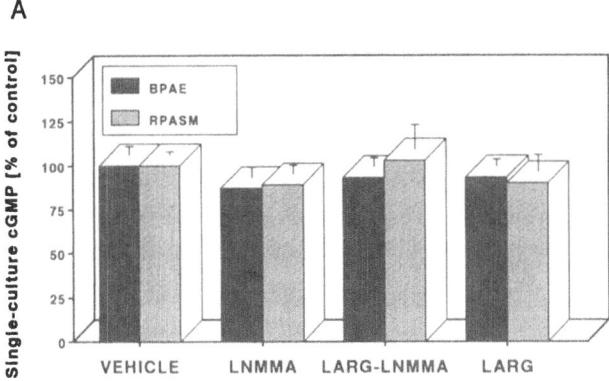

B

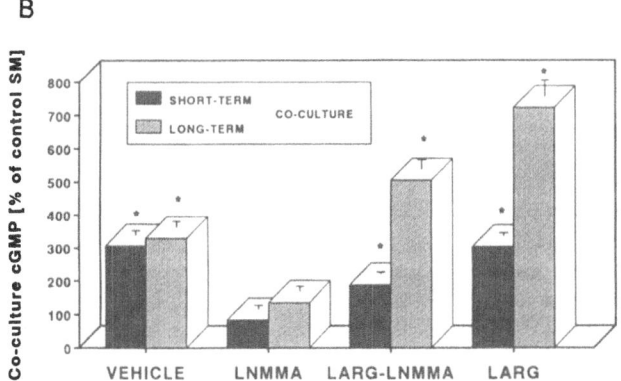

Figure 1: Effects of an inhibitor (N^G-monomethyl-L-arginine, LNMMA, 300 μM) or substrate (L-arginine, LARG, 1 mM) of endothelial synthesis of nitric oxide on the accumulation of cGMP in single cultures of bovine pulmonary arterial endothelial cells (Panel A, BPAE), in single cultures of rabbit pulmonary arterial smooth muscle cells (Panel B, RPASM) or in short-term (15 min) or long-term (3 days) co-culture of BPAE + RPASM. RPASM cells were subcultured into 24 multiwell plates and the next day endothelial cells were seeded on top of the subconfluent smooth muscle layer. Age-matched single cultures of smooth muscle and endothelial cells served as control. After 30 min incubations with L-arginine or LNMMA or both, the cells were incubated with the phosphodiesterase inhibitor 3-isobutyl-1-methyl-xanthine (IBMX, 1 mM) for 15 min and cGMP accumulation was determined by radioimmunoassay in HCl extracts. In the short-term bioassay model, endothelial cells grown on glass coverslips were selectively pretreated with L-arginine or LNMMA and after wash, the coverslips with the endothelial cells (10^5 cells/coverslip) were placed into smooth muscle cell-containing wells. After 15 minutes, the smooth muscle cells and endothelial cells were separated by removing the coverslips with the endothelial cells from the wells and cGMP levels were determined in the smooth muscle cells only.

A

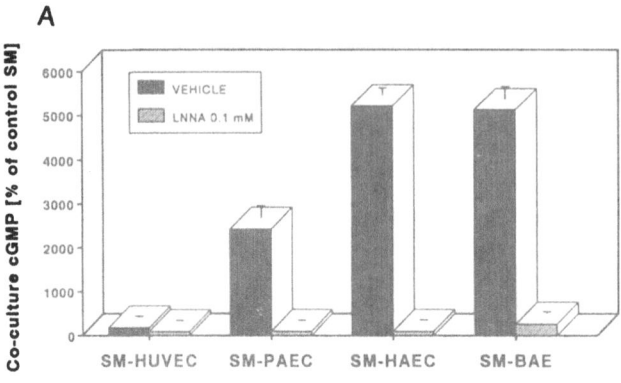

B

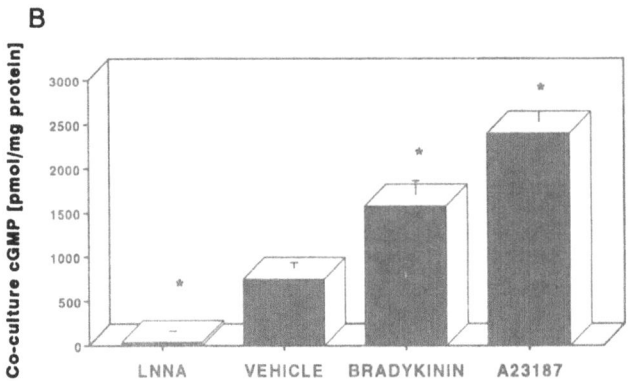

Figure 2: Modulation of smooth muscle (SM) cGMP accumulation by macrovascular endothelial cells via agonist-independent (Panel A and Panel B, VEHICLE) and bradykinin or calcium ionophore-induced release of EDRF (Panel B) in culture. HUVEC: human umbilical vein; PAEC: pig aortic, HAEC: human aortic; BA: bovine aortic endothelial cells.

METHYLENE BLUE

Methylene blue is generally assumed to inhibit the soluble isoform of guanylate cyclase, presumably by oxidizing the ferrous heme group linked to the enzyme molecule (*Ignarro et al., 1984; Martin et al., 1985*). However, methylene blue was reported to be a more potent inhibitor of the EDRF-mediated vasodilation than of relaxation to nitrovasodilators, and differences in the inhibitory actions of methylene blue were found when vascular responses to sodium nitroprusside, nitroglycerin and authentic nitric oxide were compared in the same preparation (*Ignarro et al., 1984; Martin et al., 1985; Watanabe et al., 1988; Wolin et al., 1990*). On the basis of these differences, it has been suggested that the action of methylene blue may involve superoxide anion and other reactive oxygen species to inhibit EDRF-induced vasodilation in the cerebral microcirculation of cats, and both nitric oxide- and EDRF-mediated vasodilation in the rat cremaster muscle (*Marshall et al., 1988; Wolin et al., 1990*).

In our study, methylene blue induced a dose- and time-dependent generation of superoxide anion from rabbit pulmonary arterial smooth muscle cells (RPASM), as

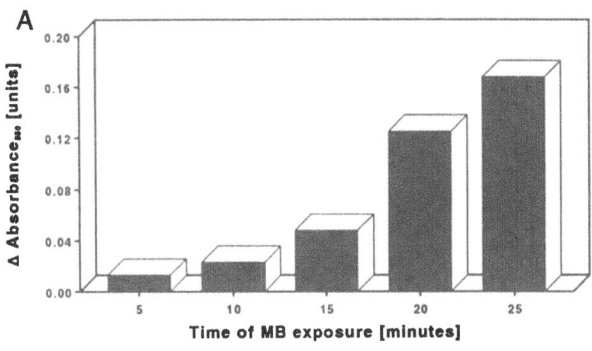

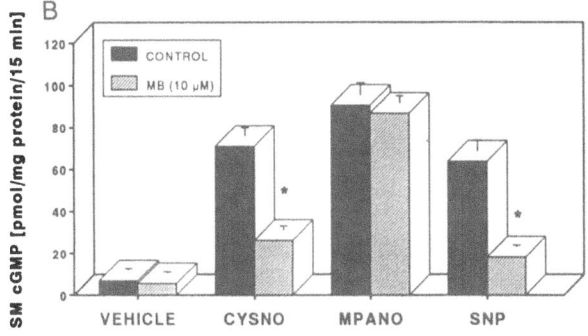

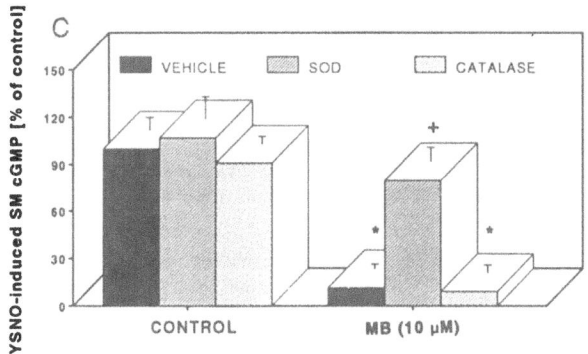

Figure 3: Role of superoxide anion in methylene-blue induced inhibition of nitrovasodilator responses. Panel A: Time-course of methylene blue (MB) on smooth muscle (SM) cell associated superoxide anion generation. Smooth muscle cells were incubated with cytochrome c alone or in the presence of methylene blue (10 μM) for the indicated time. Optical densities were determined at 550 nm and methylene blue-induced increase in absorbance was calculated. Panel B: Effects of methylene blue on cGMP accumulation in rabbit pulmonary arterial smooth muscle cells stimulated with S-nitroso-L-cysteine (CYSNO, 100 μM), S-nitroso-3-mercaptoproprionic acid (MPANO, 100 μM) and sodium nitroprusside (SNP, 100 μM). Panel C: Effects of superoxide dismutase (SOD) and catalase on methylene blue-induced inhibition of cGMP accumulation elicited by S-nitroso-L-cysteine.

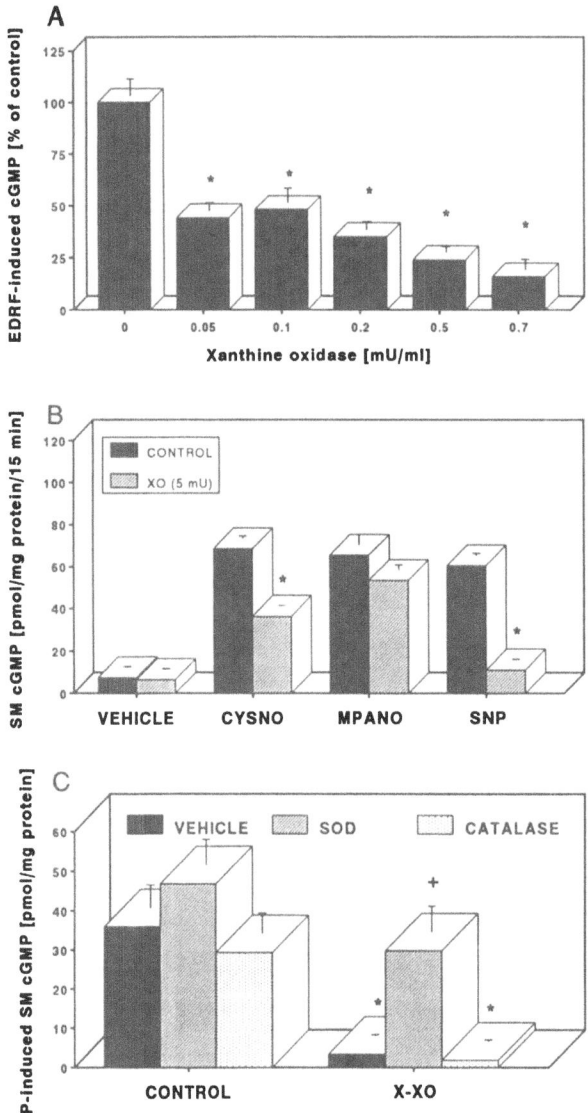

Figure 4: Role of superoxide anion in xanthine oxidase (XO)-induced inhibition of nitrovasodilator responses. Panel A: Concentration effects of XO on EDRF-induced cGMP accumulation in long-term co-culture. Panel B: Effects of XO on cGMP accumulation in rabbit pulmonary arterial smooth muscle cells stimulated with S-nitroso-L-cysteine (CYSNO, 100 μM), S-nitroso-3-mercaptoproprionic acid (MPANO, 100 μM) and sodium nitroprusside (SNP, 100 μM). Panel C: Effects of superoxide dismutase (SOD) and catalase on XO-induced inhibition of cGMP accumulation elicited by sodium nitroprusside.

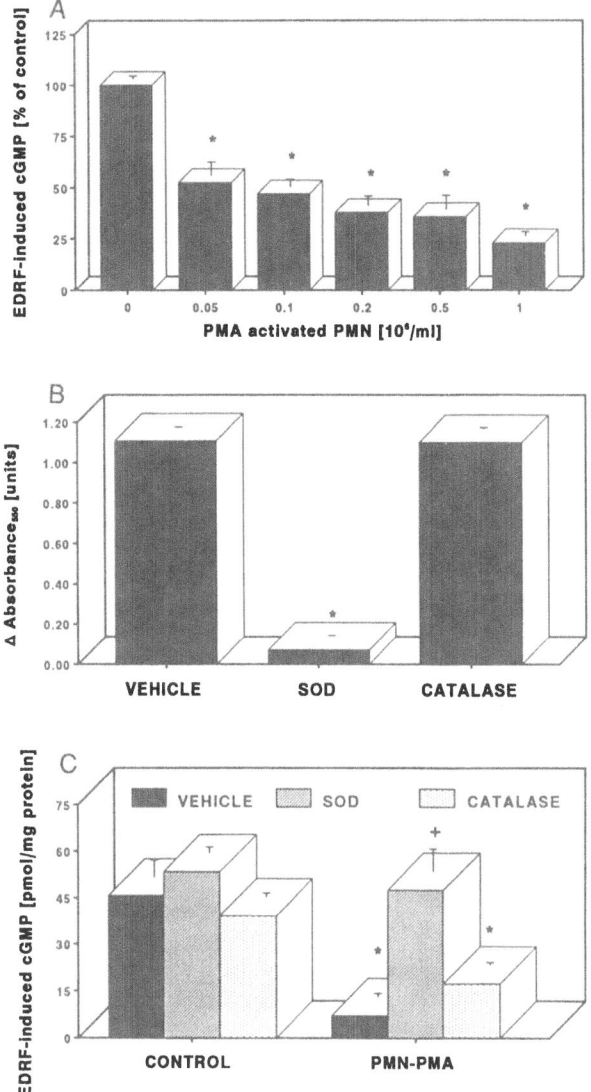

Figure 5: Role of superoxide anion in activated neutrophil-induced inhibition of EDRF-responses. Panel A: Concentration effects of PMA-activated neutrophils (PMN) on EDRF-induced cGMP accumulation in long-term co-culture. Panel B: Effects of super-oxide dismutase (SOD) and catalase on cytochrome c reduction by PMA-activated neutrophils. Panel C: Effects of superoxide dis-mutase (SOD) and catalase on neutrophil-elicited inhibition of EDRF-induced cGMP accumulation in long-term co-culture.

evidenced from spectrophotometric determination of cytochrome c reduction (Figure 3A). Associated with the generation of superoxide anion, methylene blue pretreatment inhibited S-nitroso-L-cysteine (CYSNO) and sodium nitroprusside (SNP)-induced cGMP accumulation, but responses elicited by S-nitroso-3-mercaptoproprionic acid (MPANO), the stable deaminated analogue of CYSNO, were not altered by methylene blue (Figure 3B). Furthermore, inhibition of cGMP accumulation to CYSNO and SNP

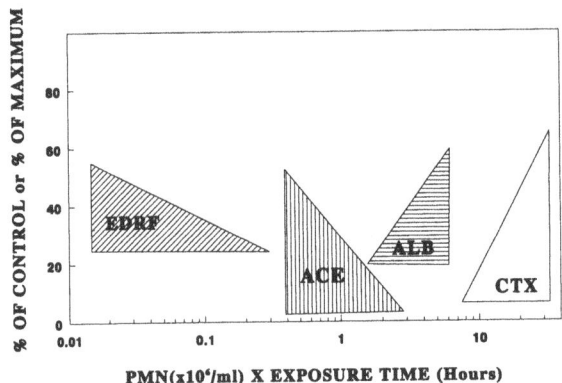

Figure 6: Comparison of neutrophil-induced EDRF dysfunction to other indexes of endothelial injury such as inhibition of angiotensin converting enzyme activity (ACE), increased permeability to albumin (ALB) and cytotoxicity (CTX) on the basis of time and neutrophil requirement to achieve injury.

by methylene blue was completely prevented by superoxide dismutase (SOD) but not catalase (Figure 3C). Interestingly, selective pretreatment of endothelial cells grown on glass coverslips with methylene blue before co-culture with untreated RPASM produced a reduction in RPASM cGMP levels of a magnitude comparable to that seen in co-cultures of methylene blue-pretreated RPASM with untreated endothelial cells, and which was partially prevented by superoxide dismutase (*Marczin et al., 1992b*). These results suggest that inactivation of soluble guanylate cyclase *per se* cannot account for the effects of methylene blue in culture, whereas, a major mechanism of inhibition appears to be superoxide inactivation of EDRF and nitric oxide released from labile nitrovasodilators.

XANTHINE OXIDASE

Xanthine oxidase (XO) has been widely used to investigate oxidant stress, because this enzyme produces reactive oxygen species, such as superoxide anion and hydrogen peroxide, during the catalytic oxidation of hypoxanthine and xanthine to uric acid. We investigated the effect of XO on EDRF-elicited cGMP accumulation in long-term co-cultures and on nitrovasodilator-induced cGMP levels in single smooth muscle cultures. Associated with extracellular generation of superoxide anion, XO inhibited EDRF action in a concentration-dependent manner (Figure 4A). The inhibition profile of XO on nitrovasodilator-induced cGMP accumulation was qualitatively similar to that elicited by methylene blue: XO had no effects on baseline cGMP levels or cGMP responses elicited by MPANO, but inhibited CYSNO and SNP responses (Figure 4B). As shown in Figure 4C, this inhibition of nitrovasodilator-induced cGMP accumulation could be prevented by the inclusion of superoxide dismutase, but not with catalase, to suggest a causative correlation between the extracellular accumulation of superoxide anion and loss of the biological activity of nitric oxide .

ACTIVATED NEUTROPHILS

Although partially reduced and activated oxygen species can be produced by numerous reactions, accumulation and activation of inflammatory cells, such as neutrophils and monocytes in the close proximity of the endothelial cells, is likely a major source of oxidant stress in various pathological conditions (*Halliwell, 1987; Fantone and Ward, 1982; Rinaldo, 1986*). To investigate the participation of EDRF in acute neutrophil-mediated endothelial injury, we studied the effects of PMA-activated

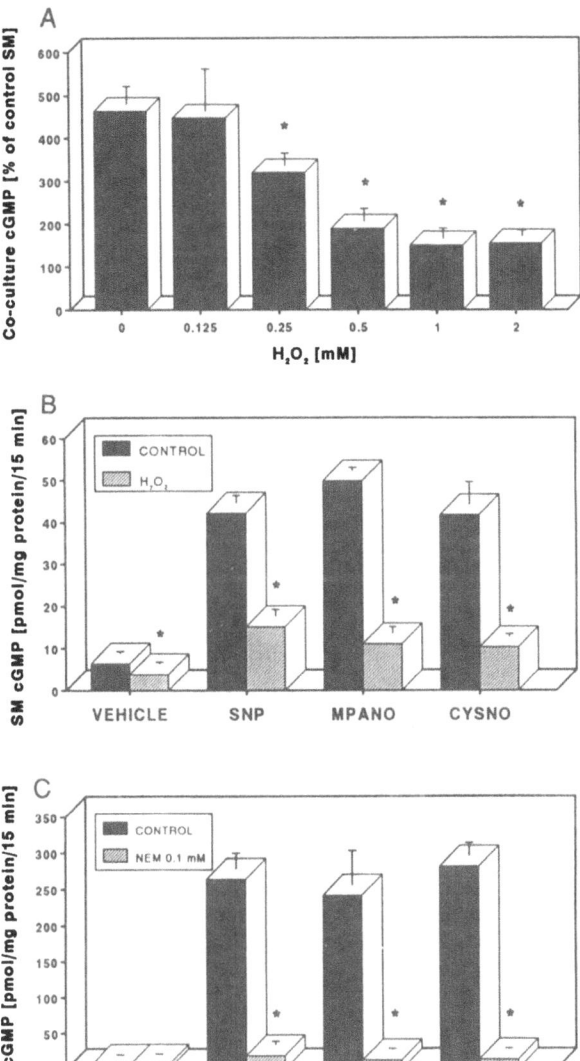

Figure 7: Hydrogen peroxide-induced inhibition of nitrovasodilator responses. Panel A: Concentration-effects of H_2O_2 on EDRF-induced cGMP accumulation. Endothelial cells were selectively exposed to increasing concentrations of H_2O_2 for 30 min prior to short-term co-culture. Panel B: Effects of H_2O_2 on cGMP accumulation in rabbit pulmonary arterial smooth muscle cells stimulated with S-nitroso-L-cysteine (CYSNO, 100 μM), S-nitroso-3-mercaptoproprionic acid (MPANO, 100 μM) and sodium nitroprusside (SNP, 100 μM). Panel C: Effects of N-ethyl-maleimide (NEM), a thiol depleting agent on nitrovasodilator-stimulated cGMP accumulation.

neutrophils on EDRF-induced cGMP accumulation, as well as on activity of angiotensin converting enzyme, endothelial monolayer permeability to albumin and release of [51]chromium, as a measure of cytotoxicity. Nonactivated neutrophils had no effect upon smooth muscle basal cGMP, suggesting the absence of detectable amounts of constitutively released NO from these neutrophils. Nonstimulated neutrophils generated small amounts of superoxide anion, and accordingly, up to 1 million/well did not alter SNP and EDRF-mediated cGMP accumulation. However, activation of increasing numbers of neutrophils with PMA (10 ng/ml) resulted in a concentration-dependent inhibition of EDRF-elicited cGMP responses (Figure 5A). Activation of only 100 neutrophils/mm^3 medium resulted in nearly 50 percent loss of the biological activity of basally released EDRF. At a constant neutrophil concentration (1 million/ml), the impairment of EDRF activity was dependent on the concentration of PMA, thus correlated with the level of activation of neutrophils. This effect of neutrophils was rapid, required only 20 min co-incubation with the endothelial-smooth muscle co-culture and was associated with generation of superoxide anion as gauged from superoxide dismutase-but not catalase-inhibitable reduction of cytochrome c (Figure 5B). SOD not only inhibited cytochrome c reduction, but it reversed the inhibition of cGMP accumulation as well, suggesting superoxide as a predominant species responsible for the loss of the EDRF activity under these experimental conditions (Figure 5C). Interestingly, catalase also tended to restore cGMP levels, suggesting that hydrogen peroxide (although to much less extent than superoxide) might also be involved. Comparison of the time-course and concentration-dependence of EDRF-inhibition to other indexes of endothelial injury revealed that ACE-inhibition and changes in endothelial permeability occurred after much longer (4h) incubation times with similar concentrations of neutrophils, whereas cytotoxicity was a late event (4-8h) and required more neutrophils (1-4 x 10^6/ml, Figure 6).

HYDROGEN PEROXIDE-INDUCED INHIBITION OF EDRF ACTION

Beside superoxide anion, hydrogen peroxide has also been implicated in diverse forms of vascular injury, however its effect on vasoreactivity is highly controversial (*Weiss et al, 1981*). Depending on the vascular bed and basal tone of the vessel, H$_2$O$_2$ might be either a vasodilator or a constrictor. Some of these actions are presumably direct effects on smooth muscle cells, however in the cat cerebral microcirculation, H$_2$O$_2$ specifically eliminated EDRF release in response to acetylcholine (*Wei and Kontos, 1990*). To obtain more information about potential interactions between H$_2$O$_2$ and the EDRF pathway, we investigated cGMP accumulation in response to various nitrovasodilators and EDRF released basally from endothelial cells.

Pretreatment of long-term co-cultures with H$_2$O$_2$ for 30 min resulted in a dose-dependent decrease in cGMP formation (49-79%), suggesting that either the endothelial EDRF synthesis or transport or smooth muscle responsiveness was altered by the oxidant. The short-term co-culture model offered the opportunity to selectively treat either only the endothelial or smooth muscle cells prior to establishment of the co-culture. In short-term co-cultures of H$_2$O$_2$-pretreated endothelial cells with untreated smooth muscle cells, cGMP levels were reduced suggesting the inability of endothelial cells to generate EDRF (Figure 7A). The action of H$_2$O$_2$ on endothelial cells was mimicked by thiol-depleting agents such as N-ethyl-maleimide. Pretreatment of smooth muscle cells with H$_2$O$_2$, however, also attenuated EDRF-induced cGMP accumulation to suggest that smooth muscle cell responsiveness to EDRF was also altered. Importantly, under these conditions general cellular morphology and ^{51}Cr release from prelabelled cells were unchanged and isoproterenol and IBMX-elicited cAMP accumulation was only marginally affected. These data suggest a rather specific effect of H$_2$O$_2$ on the cGMP pathway (*Marczin et al., 1992c*). Hydrogen peroxide also inhibited cGMP increases in response to CYSNO and SNP, and unlike methylene blue and xanthine oxidase, it also reduced cGMP formation induced by the stable nitrosothiol MPANO (Figure 7B). These actions of H$_2$O$_2$ could be prevented by pretreatment with either dimethylthiourea, deferoxamine or dithiotreitol suggesting a mechanism of toxicity involving iron-catalyzed formation of intracellular hydroxyl radicals and their attack on cellular thiols. H$_2$O$_2$ oxidized both protein and non-protein

12

thiols in smooth muscle cells, and its action was mimicked by a thiol depleting agent N-ethyl-maleimide (Figure 7C).

SUMMARY AND IMPLICATIONS

We have investigated the fate of agonist-independent EDRF release under a variety of oxidant stress conditions. From these studies and the observations of others, it is evident that the physiologically relevant, homeostatic communication between vascular endothelial and smooth muscle cells via basally released L-arginine-derived EDRF is maintained in culture and can be monitored (although indirectly) at a molecular level with simple and sensitive assay for cGMP. The factors responsible for this agonist independent release of EDRF are not clear yet, but basal activity of the cNOS and some mechanical stimuli associated with handling of the cells might be involved. Similarly to agonist-induced release of EDRF, oxidant stress produces profound changes in the synthesis, transport and action of basally released EDRF, generally resulting in a decreased biological activity of EDRF. Our study suggests three major interactions between EDRF and oxidants. One, and probably the most common reaction, is the superoxide anion-induced direct inactivation of EDRF, as occurred with methylene blue, xanthine oxidase and during the early phase of neutrophil-mediated injury. In some conditions, this interaction might be localized in the extracellular space or at the surface of the endothelial cells or intracellular compartments. These microenvironments might represent a varying accessibility for different agents and might be a limiting factor for their therapeutic potential. In this regard, the use of small molecular weight superoxide anion scavengers or mimics might prove to be more useful than the large and charged superoxide dismutase. The interplay of EDRF and superoxide not only results in decreased biological activity of EDRF but also might eliminate toxic activity of the superoxide anion or actually give rise to peroxynitrite, a more toxic reactive species (*Feigl, 1989; Rubanyi, 1988, Beckman et al., 1990*). The question of how these interactions relate to the ultimate expression of tissue injury still remains to be determined.

Another mechanism whereby EDRF action might be limited is the interference with EDRF generation and release from the endothelial cells. This might involve inactivation of the cNOS and reduction of cofactors necessary for EDRF synthesis. Our data, and those reported by others, suggest an important role for protein and nonprotein thiols for the full expression of EDRF biological activity (*Hecker et al., 1992; Minor et al., 1989*). These thiols are ready subjects for oxidation by hydrogen peroxide or hydroxyl radicals with the consequences of reduced EDRF generating activity.

Similar mechanism might be responsible for reduced responsiveness of smooth muscle cells to basally released EDRF. Needleman and Ignarro suggested that thiols located within the smooth muscle are necessary to bring about vasodilation or sGC stimulation by certain nitrovasodilators (*Needleman et al., 1973; Ignarro et al., 1981*). Our data suggest that cGMP formation in response to basally released EDRF requires a critical thiol, either at the sGC itself, or at a yet unidentified step in the cascade of EDRF "metabolism" within the smooth muscle cells.

The clinical relevance of these concepts is continuously evolving. More and more data appear to suggest that oxidant stress is one major mechanism underlying endothelial dysfunction in clinical settings. It is likely that interference with basally released EDRF is a relatively early phenomenon during oxidant injury and, depending on the specific reaction involved, a variety of therapeutic possibilities can be considered to protect this important regulatory mechanism.

ACKNOWLEDGEMENTS

We are pleased to acknowledge the expert technical assistance of Ms. Connie Snead, Ms. Livia Jozsa Marczin and Mr. Jim Parkerson. Supported by HL31422.

REFERENCES

Aisaka, K., Gross, S.S., Griffith, O.W., and Levi, R., 1989, N^G-Methylarginine, an inhibitor of endothelium- derived nitric oxide synthesis, is a potent pressor agent in the guinea-pig: Does nitric oxide regulate blood pressure *in vivo*? *Biochem. Biophys. Res. Commun.* 160: 881-886.

Bassenge, E., 1991, Endothelium-mediated regulation of coronary tone. *Basic Res. Cardiopl.* 86 (Suppl.2):69-76.

Beckman, J.S., Beckman, T.W., Chen, J., Marshall, P.A., and Freeman, B.A., 1990, Apparent hydroxyl radical production by peroxynitrite: Implications for endothelial injury from nitric oxide and superoxide anion. *Proc. Natl. Acad. Sci.* 87: 1620-1624.

Bredt, D.S., Hwang, P.M., Glatt, C.G., Lawenstein, C., Reed, R.R., and Synder, S.H., 1991, Cloned and expressed nitric oxide synthase structurally resembles cytochrome P-450 reductase. *Nature* 351:714-718.

Busse, R., Trogish, G., and Bassenge, E., 1985, The role of the endothelium in the control of vascular tone. *Basic Res. Cardiol.* 80:475-490.

Chu, A., Chambers, D.E., Lin, C.C, Kuehl, W.D. and Palmer, R.M.J., 1991, Effects of inhibition of nitric oxide formation on basal vasomotion and endothelium-dependent responses of the coronary arteries in awake dogs. *J. Clin. Invest.* 87:1964-1968.

Fantone, J.C., Ward, P.A., 1982, Role of oxygen-derived free radicals and metabolites in leukocyte-dependent inflammatory reactions. *Am. J. Pathol.* 107:394-418.

Feigl, E.O., 1989, EDRF: a protective factor? *Nature* 331:490-491.

Förstermann, U., Schmidt, H.H.H.W., Pollock, J.S., Sheng, H., Mitchell, J.A., Warner, T.D., Nakane, M. and Murad, F., 1991, Isoforms of EDRF/NO synthase. Characterization and purification from two different cell types. *Biochem. Pharmacol.* 41:1849-1857.

Furchgott, R.F., 1988, Studies on relaxation of rabbit aorta by sodium nitrate: basis for the proposal that the acid-activatable component of the inhibitory factor from retractor penis is inorganic nitrate and the endothelium-derived relaxing factor is nitric oxide. In *Mechanisms of Vasodilation* (Vanhoutte, P.M. ed) pp.401-414, Raven Press, New York.

Furchgott, R.F. and Zawadzki, J.V., 1980, The obligatory role of endothelial cells in the relaxation of arterial smooth muscle by acetylcholine. *Nature* 288:373-376.

Ganz, P., Davies, P.F., Leopold, J.A., Gimbrone, M.A., and Alexander, R.W., 1986, Short- and long-term interactions of endothelium and vascular smooth muscle in co-culture:effects on cyclic GMP production. *Proc. Natl. Acad. Sci. USA.* 83:3552-3556.

Gardiner, S.M., Compton, A.M., Bennett, T., R.M.J. Palmer, and Moncada, S., 1990, Control of regional blood flow by endothelium-derived nitric oxide. *Hypertension.* 15:486-492.

Gryglewski, R.J., Palmer, R.M.J., and Moncada, S., 1986, Superoxide anion is involved in the breakdown of endothelium-derived vascular relaxing factor. *Nature.* 320:454-456.

Halliwell, B., 1987, Oxidants and human disease: some new concepts. *FASEB J* 1:358-364.

Hecker, M., I. Siegle, H., Macarthur, W.C., Sessa, and Vane, J.R., 1992, Role of intracellular thiols in release of EDRF from cultured endothelial cells. *Am. J. Physiol.* 262 (*Heart Circ.Physiol.*31): H888-H8968.

Ignarro, L.J., Byrns, R.E., and Wood, K.S., 1988, Biochemical and pharmacological properties of endothelium-derived relaxing factor and its similarity to nitric oxide radical. In *Mechanisms of Vasodilation* (Vanhoutte, P.M. ed) pp.427-435, Raven Press, New York.

Ignarro, L.J., Burke, T.M., Wood, K.S., Wolin, M.S., Kadowitz, P.J., 1984, Association between cyclic GMP accumulation and acetylcholine-elicited relaxation of bovine intrapulmonary artery. *J. Pharmacol. Exp. Ther.* 228:682-690.

Ignarro, L.J., H. Lippton, J.C. Edwards, W.H. Baricos, A.L. Hyman, P. J. Kadowitz and C.A. Gruetter, 1981, Mechanism of vascular smooth muscle relaxation by organic nitrates, nitrites, nitro-prusside and nitric oxide: evidence for the involvement of S-nitrosothiols as active intermediates. *J. Pharmacol. Exp. Ther.* 218:739-749.

Loeb, A.L., Johns, R.A., Milner, P., and Peach, M.J., 1985, Endothelium-derived relaxing factor in cultured cells. *Hypertension* 7:804-807.

Lucchesi, B.R., Mullane, K.M., 1986, Leukocytes and ischemia-induced myocardial injury. *Annual Review of Pharmacology and Toxicology* 26:201-224.

Marczin, N., Ryan, U.S., and Catravas, J.D., 1992a, Endothelial cGMP does not regulate basal release of endothelium-derived relaxing factor in culture. *Am. J. Physiol.* 263 *Lung cell. Mol. Physiol.* 7):L113-L121.

Marczin, N., Ryan, U.S., and Catravas, J.D., 1992b, Methylene blue inhibits nitrovasodilator- and EDRF-induced cGMP accumulation in cultured pulmonary arterial smooth muscle cells via generation of superoxide anion. *J. Pharmacol. Exp. Ther.* 263: 170-179.

Marczin, N., Ryan, U.S., and Catravas, J.D., 1992c, Effects of oxidant stress on endothelium-derived relaxing factor-induced and nitrovasodilator-induced cGMP accumulation in vascular cells, in culture. *Circ. Res.* 70:326-340.

Marshall, J.J., Wei, E. P. and Kontos, H. A., 1988, Independent blockade of cerebral vasodilation from acetylcholine and nitric oxide. *Am. J. Physiol.* 255: H847-H854.

Marshall, J.J. Kontos, H.A., 1990, Endothelium-derived relaxing factors: a perspective from *in vivo* data. *Hypertension* 16:371-386.

Martin, W., Villani, G.M., Jothianandan, D., and Furchgott, R.F., 1985, Selective blockade of endothelium dependent and glyceryl trinitrate-induced relaxation by hemoglobin and by methylene blue in the rabbit aorta. *J. Pharmacol. Exp. Ther.* 232: 708-716.

May, G.R., Crook, P., Moore, P.K., and Page, C.P., 1991, The role of nitric oxide as an endogenous regulator of platelet and neutrophil activation within the pulmonatory circulation of the rabbit. *Br. J. Pharmacol.* 102: 759-763.

Minor, R.L., Myers, P.R., Bates, J.N., and Harrison, D.G., 1989, Basal EDRF release is reduced by depletion of cystein from endothelial cells (Abstract). *Circulation* 80, Suppl. 2:1121.

Myers, P.R., Guerra, R. Jr., Harrison, D.G., 1989, Release of NO and EDRF from cultured bovine aortic endothelial cells. *Am. J. Physiol.* 256:H1030-H1037.

Myers, P.R., Minor, R.L., Guerra, R., Bates, J.N. and Harrison, D.G., 1990, Vasorelaxant properties of the endothelium-derived relaxing factor more closely resemble S-nitrosocysteine than nitric oxide. *Nature* 345:161-163.

Needleman, P., Jakschik, B., and Johnson, E.M., 1973, Sulfhydryl requirement for relaxation of vascular smooth muscle. *J. Pharmacol. Exp. Ther.* 187:324-331.

Palmer, R.M.J., Ferroge. A.G., and Moncada, S., 1987, Nitric oxide release accounts for the biological activity of endothelium-derived relaxing factor. *Nature* 327:524-526.

Palmer, R.M.J., Ashton, D.S., and Moncada, S., 1988, Vascular endothelial cells synthesize nitric oxide from L-arginine. *Nature* 333:664-666.

Radomski, M.W., Palmer, R.M.J., and Moncada, S., 1987, The role of nitric oxide and cGMP in platelet adhesion to vascular endothelium. *Lancet.* 2: 1057-1058.

Rees, D.D., Palmer, R.M.J., and Moncada, S., 1989, Role of endothelium-derived nitric oxide in the regulation of blood pressure. *Proc. Natl. Acad. Sci. USA* 86:3375-3378.

Rinaldo, J.E., 1986, Mediation of ARDS by leukocytes: clinical evidence and implications for therapy. *Chest* 89:590-593.

Rubanyi, G.M., 1988, Potential role of endothelium-derived relaxing factor in the protection against free radical injury. *J. Mol. Cell. Cardiol.* 20 (suppl. V): 2S56.

Rubanyi, G.M., and Vanhoutte, P.M., 1986, Superoxide anion and hyperoxia inactivate endothelium-derived relaxing factor. *Am. J. Physiol.* 250, H822-H827.

Rubanyi, G.M., Romero, J.C., and Vanhoutte, P.M., 1986, Flow-induced release of endothelium-derived relaxing factor. *Am. J. Physiol.* 250, H1145-H1149.

Savitsky , J.P., Doczi, J., Black, J., Arnold, J.D., 1978, A clinical safety trial of stroma-free hemoglobin. *Clin. Pharmacol. Ther.* 23:73-80.

Schmidt, H.H.H.W., Zernikov, B., Baeblich, S., Böhme, E., 1990, Basal and stimulated formation and release of L-arginine-derived nitrogen oxides from cultured endothelial cells. *J. Pharmacol. Exp. Ther.* 254:591-597.

Tolins, J., Palmer, R.M.J., Moncada, S., and Raij, L., 1990, Role of endothelium-derived relaxing factor in regulation of renal hemodynamic responses. *Am. J. Physiol.* 258:H655-H772.

Vallance, P.J., Collier, J., and Moncada, S., 1989, Effects of endothelium-derived nitric oxide on peripheral arteriolar tone in man. *Lancet.* 2:997-1000.

Watanabe, M., Rosenblum, W.I. and Nelson, G.H., 1988, *In vivo* effect of methylene blue on endothelium-dependent and endothelium-independent dilations of brain microvessels in mice. *Circ. Res.* 62:86-90.

Wei, E.P., Kontos, H.A., 1990, H_2O_2 and endothelium-dependent cerebral arteriolar dilation *Hypertension* 16:162-169.

Wei, E.P., Kontos, H.A., Christman, C.W., DeWitt, D.S. and Povlishock, J.T., 1985, Superoxide generation and reversal of acetylcholine-induced cerebral arteriolar dilation after acute hypertension. *Circ. Res.* 57:781-78.

Weiss, S.J., Young, J., LoBuglio, A.F., Slivka, A., 1981, Role of hydrogen peroxide in neutrophil-mediated destruction of cultured endothelial cells. *J. Clin. Invest.* 68:714-72.

Wiklund, N.P., Persson, M.G., Gustafsson, L.E., Moncada, S., and Hedqvist, P., 1990, Modulatory role of endogenous nitric oxide in pulmonary circulation *in vivo*. *Eur. J. Pharm.* 185:123-124.

Wolin, M.S., Cherry, P.D., Rodenburg, J.M., Messina, E.J. and Kaley, G., 1990, Methylene blue inhibits vasodilation of skeletal muscle arterioles to acetylcholine and nitric oxide via the generation of superoxide anion. *J. Pharmacol. Exp. Ther.* 254: 872-876.

II. INVASION AND NEOPLASIA

INTERACTION OF MALARIA-INFECTED

CELLS WITH THE VASCULAR WALL

Giorgio Senaldi, Fabienne Tacchini-Cottier
and Georges E. Grau

WHO-Immunology Research and Training Centre
Department of Pathology
University of Geneva
Switzerland

INTRODUCTION

Recent estimations confirm malaria as the most common infectious disease affecting the human species. Worldwide, more than 2 billion people are at risk of infection. About 200 million cases occur every year and approximately 2 million of them result in the death of the patient, especially children in endemic areas (*Greenwood et al., 1987*). *Plasmodium falciparum* infection is by far the most severe form of malaria, which accounts for the vast majority of fatal cases. It owes its severity to its frequency of complications (*White, 1986*). The most dangerous complication is cerebral malaria (CM), which is responsible for about 80% of all fatal cases, although it develops in only 0.5-1% of the episodes of *P. falciparum* infection. CM is invariably lethal if untreated, and it also kills up to 40% of treated patients. The recovery from CM is occasionally accompanied by permanent, disabling neurological sequelae (*Marsh and Greenwood, 1986; Phillips and Warrell, 1986; World Health Organization Malaria Action Program, 1986; Warrell, 1987*).

Central to the pathogenesis of CM is the formation of a cellular plug in brain capillaries and post-capillary venules, subsequently entailing vessel occlusion, ischaemia, tissue damage - manifested first by necrosis of the vascular wall and peri-vascular hemorrhages - and functional impairment (*Aikawa et al., 1990; Roman, 1991*). The cells involved in the formation of the vessel plug are chiefly red blood cells (RBC), especially those which are parasitized (PRBC); white blood cells (WBC) and platelets are also present in the vessel plug but to a much lesser extent (*MacPherson et al., 1985*).

P. berghei ANKA infection provides in some strains of mice an experimental model of malaria which has strong tendency to get complicated by CM (*Grau et al., 1989a*). The formation of a cellular plug in brain capillaries and post-capillary venules is also critical as to the pathogenesis of this type of CM (*Grau et al., 1989a*). In contrast to human CM, however, the cells chiefly involved in the formation of the plug are mononuclear WBC, whereas RBC are less frequently visible and granulocytes are rarely encountered (*Grau et al., 1989a*). That the immune system, and tumor necrosis factor (TNF) in particular, play a pathogenetic role in human CM was suggested by studies performed on this model of disease.

It is the aim of this paper to review: 1) how *P. falciparum* infection confers stickiness on PRBC and increased adhesiveness on the endothelium of the brain microcirculation (see footnote), which results in cytoadherence and vessel plug

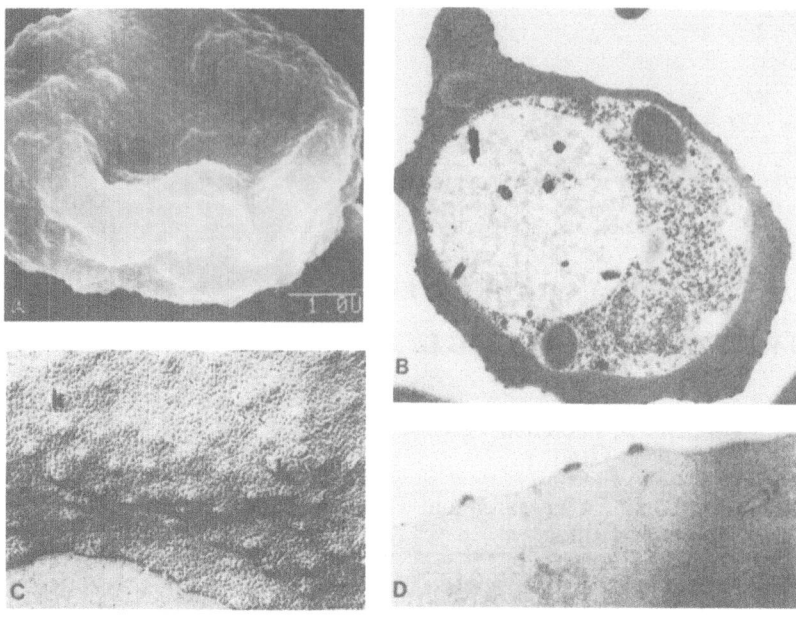

Figure 1: Knobs on the surface of a *P. falciparum*-parasitized red blood cell. A and C: scanning electron microscopy photographs. B and D: transmission electron microscopy photographs. Knobs are electron-dense conoid protrusions of the PRBC membrane (*reproduced from Sherman et al., 1992*).

formation; and 2) how *P. berghei* ANKA infection implicates enhanced mononuclear WBC stickiness and adhesiveness of the brain microcirculation endothelium, hence cytoadherence and, again, vessel plug formation. A common physiopathological pathway leading to vessel plug formation in human and murine CM will be illustrated, together with the rationale for an anti-adhesion intervention to prevent and treat CM.

CYTOADHERENCE IN HUMAN CM

This phenomenon is eminently the result of both stickiness acquired by the PRBC, which is a consequence of modifications conferred on the RBC by *P. falciparum* upon invasion, and adhesiveness, i.e., upregulation of the expression of adhesion molecules, displayed by the endothelium in the brain microcirculation, which is induced by cytokines, especially TNF, produced as part of the immune response against the parasite.

PRBC stickiness: physical, morphological and molecular components. Within the RBC *P. falciparum* undergoes a 48 hour cycle of asexual reproduction, during which the merozoite that initially invaded the RBC becomes a ring first, then a trophozoite and a schizont later. By the division of the schizont and PRBC rupture, a new generation of invasive merozoites is engendered and spread in the bloodstream which continues the infection (*Wernsdorfer, 1980*). Plasmodial growth is accompanied by important physical, morphological and molecular modifications of the RBC (*Sherman et al., 1992*), some of which render this cell unsuitable for optimal flowing through the microcirculation and abnormally sticky to the vascular wall.

The physical changes which the RBC acquires following infection are gross modifications in size and shape and loss of density (*Gruenberg et al., 1983; Nash et al., 1989*). These and a reduction of the cholesterol content in the PRBC membrane (*Maguire and Sherman, 1990*) imply decreased deformability of the PRBC and compromise its rheological properties. The increase in size and the alteration in shape

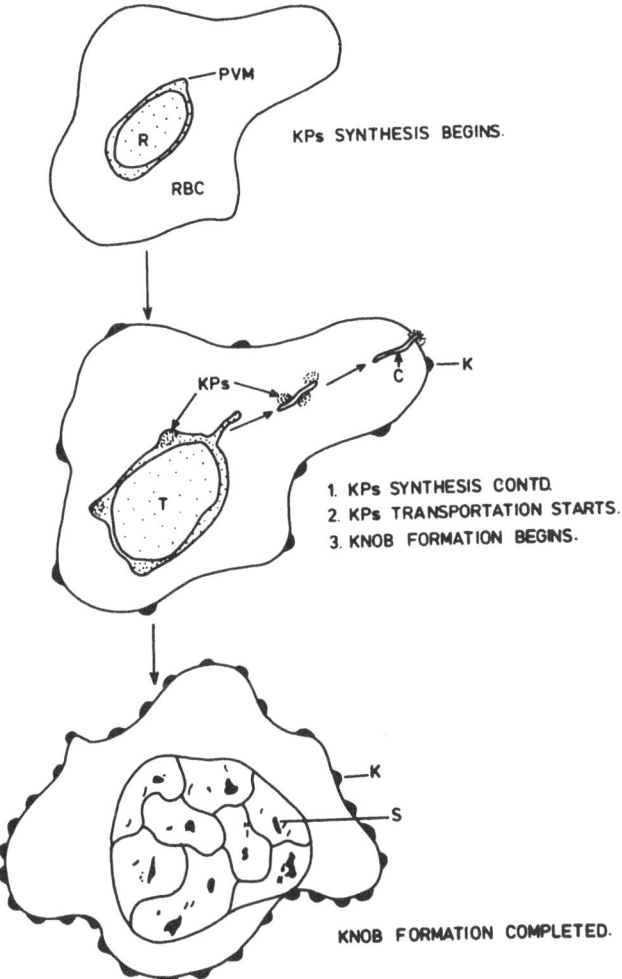

Figure 2: Knob formation and intra-erythrocytic plasmodial growth. Knob formation parallels the development of P. falciparum - here illustrated from the ring to the schizont stage - inside the PRBC. Knob proteins start being synthesised by the parasite at the ring stage, they are then transported to the membrane via the so-called Maurer's clefts (which like knobs are electron-dense), and result in full knob formation at schizont stage. KP: knob protein. PVM: parasitophorous vacuole membrane. R: ring. RBC: red blood cell. K: knob. C: Maurer's cleft. T: trophozoite. S: schizont (reproduced from Sharma, 1991).

of the PRBC, which mainly consists in the loss of the biconcave profile of the normal erythrocyte (*Gruenberg et al., 1983*), are directly related to the inner presence of the parasite, whose body size, at the trophozoite and schizont stages, is considerable (*Aikawa and Atkinson, 1990*). Because of the high content in water of the plasmodial body, which is higher than the one of the RBC, infected erythrocytes are less dense (*Nash et al., 1989*). *P. falciparum* PRBC show reduced levels of membrane cholesterol, probably as a consequence of parasite catabolic action (*Maguire and Sherman, 1990*). Low cholesterol in the RBC membrane ultimately determines an increase in membrane stiffness and loss of deformability of the whole cell. The poor deformability of PRBC that follows parasite development does not directly imply PRBC stickiness;

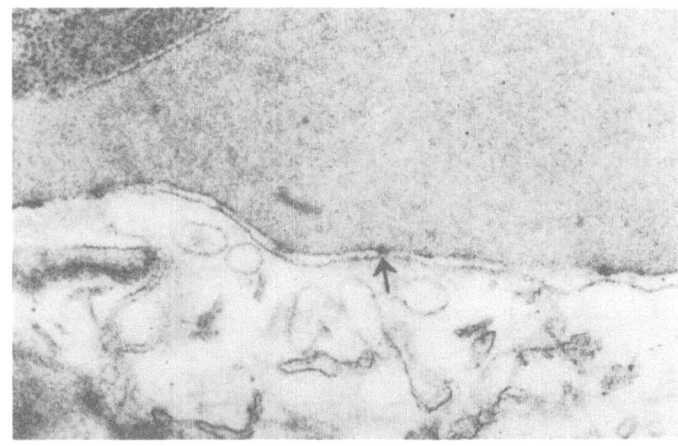

Figure 3: Knob mediated cytoadherence of a parasitized red blood cell to the brain microcirculation endothelium. Knobs (arrow) are critical sticky centers on PRBC surface and the contact points between the PRBC and the brain microcirculation endothelium, as shown in this transmission electron microscopy photograph (*reproduced from Aikawa et al., 1990*).

however, it alters RBC velocity and provides for prolonged contact of the PRBC with the endothelium, thus favoring cytoadherence.

The morphological change that markedly characterizes *P. falciparum* PRBC and is central to its acquired stickiness is the appearance of knobs on its surface (*Sharma, 1991*). Knobs are conoid and electron-dense protrusions (diameter 110-160 nm; height 30-40 nm) of the erythrocyte membrane (Fig. 1), which appear approximately 24 hours after RBC invasion and gradually increase in number, so that as much as 5% of the surface of a PRBC can be covered by knobs at the moment that a schizont has reached maturity (Fig. 2) (*Trager et al., 1966; Langreth et al., 1978; Gruenberg et al., 1983*). Knobs are the result of a dynamic biosynthetic program carried out by *P. falciparum* during its intraerythrocytic growth. The critical role played by knobs in mediating cytoadherence of PRBC to the endothelium has originally been documented by ultrastructural studies (Fig. 3) (*Oo et al., 1987; Aikawa, 1988; Aikawa et al., 1990*). Although stickiness is displayed by the PRBC membrane also outside knobs - as it is shown by the existence of knobless yet cytoadherent PRBC - these remain the specialized attachment points on the PRBC to the endothelium, where novel adhesion molecules produced by the parasite are either uniquely expressed - as it seems to be the case for the thrombospondin receptor (see below) - or concentrated and structurally organized.

The molecular changes that develop in a PRBC depend on the production of parasite-encoded proteins and on the modifications induced by the parasite of normal, constitutive RBC proteins. A description is given hereby of some of these novel proteins, whose list is continuously growing, which have been shown to be involved in conferring stickiness on the PRBC, by either directly mediating its adhesion to the endothelium or by producing knobs on its surface, which in turn promote stickiness (*Howard and Gilladoga, 1989; Hommel, 1990; Howard et al., 1990; Sharma, 1991; Sherman et al., 1992*).

Ockenhouse *et al* (*1991*) identified a *P. falciparum*-encoded protein on PRBC surface which directly mediates adhesion to the endothelium. These investigators, with an elegant approach, made use of a rabbit antiserum raised against the idiotype of the murine monoclonal antibody OKM8, which is directed against human CD36 (*Talle et al., 1983*), an adhesion molecule that· they previously found to be involved in the cytoadherence of the PRBC to the endothelium (see below). Such an antiserum stained PRBC in an immunofluorescence assay and blocked PRBC adhesion to melanoma cells in a test of cytoadherence (*Ockenhouse et al., 1991*). Moreover, it precipitated a radioactive molecule from I^{125} surface labelled PRBC (*Ockenhouse et al., 1991*). This molecule, termed sequestrin (*Ockenhouse et al., 1991*), is a protein of molecular weight (MW) of approximately 270 kDa, and it is the recognition molecule for CD36 on endothelial cell (EC) surface. CD36 "image" was retained by Ockenhouse and colleagues' antiserum. It is a component of the outer PRBC membrane and it is knob-associated on knobby PRBC, whereas it is spread in isolated form in the plasma membrane on knobless PRBC (*Ockenhouse et al., 1991*).

Howard *et al* (*1988*) recognized an intriguing *P. falciparum*-encoded protein on PRBC surface, which is implicated in both cytoadherence and antigenic diversity. To the identification of this protein led studies, reviewed by Howard and Gilladoga, (*1989*) and by Howard *et al.* (*1990*), which were prompted by the observation that adhesion of PRBC to the endothelium can be blocked by treatment of these cells with proteases (*Leech et al., 1984*) or with immune sera in a strain-specific fashion (*Udeinya et al., 1983*). This protein, named erythrocyte membrane protein (EMP) 1, has a MW of about 300 kDa, and it directly mediates adhesion to the endothelium (*Howard et al., 1988*), although no cognate ligand is known for it on the EC. EMP 1 is a resident of the outer PRBC membrane and it is not knob-associated, as shown by electron microscopy observations (*Aikawa and Atkinson, 1990*). Howard *et al.* have suggested that EMP 1 is a chimeric protein with an invariant portion involved in cytoadherence and a variable one responsible for antigenic diversity (*Howard et al., 1990*).

Kilejian and Jensen (*1977*) identified the first *P. falciparum*-encoded protein in PRBC and called it "knob protein", since they attributed to it the function of knob formation. Recently, given its primary structure, it has been renamed histidine-rich protein (HRP) 1 (*Howard et al., 1986*). HRP 1 has a MW of 90 kDa, and it does not directly mediate the contact between the PRBC and the EC (*Howard et al., 1986; Taylor et al., 1987*). Indeed, HRP 1 is localized in the inner PRBC membrane, where it is found concentrated at the base of knobs in addition to being abundantly present in the PRBC cytoplasm (*Taylor et al., 1987; Aikawa and Atkinson, 1990*). However, since it is necessary to knob formation, probably because it organizes knob structure, and since knobs are critical centers of PRBC stickiness, HRP 1 has to be considered an essential ingredient of this property.

EMP 2 (*Howard et al., 1987*), HRP 2 (*Howard et al., 1986*), protein 11.1 (*Petersen et al., 1990*) and the ring-parasitized erythrocyte surface antigen (RESA) (*Culvenor et al., 1991*) are other *P. falciparum*-encoded proteins, which are not related to PRBC stickiness. Involved in conferring stickiness upon PRBC and possibly encoded by *P. falciparum* is the antigen targeted by the human monoclonal antibody 33G2, prepared and characterized by Udomsangpetch *et al* (*1989a and 1989b*), which has the property to inhibit the adhesion of PRBC to melanoma cells. However, such an antigen remains ill-identified. Although there exist data which suggest that it is a protein of MW similar to the one of EMP 1, the evidence available so far indicates that 33G2 recognizes a cross-reactive epitope present on EMP 1, protein 11.1 and RESA (*Ahlborg et al., 1991*). The ability of 33G2 to block cytoadherence might therefore be due to its specificity for EMP 1.

The observations in freeze fracture electron microscopy of alterations in the distribution of intra-membranous particles in PRBC, which arise paralleling the intra-erythrocytic growth of the parasite and support knob formation, indicated the possibility that *P. falciparum* modifies band-3 (*Aikawa et al., 1985; Allred et al., 1986*), the structural protein that as monomer, dimer or tetramer constitutes the intra-membranous particles of a normal RBC (*Low, 1986; Ruoslahti and Giancotti, 1989*). Indeed, different modifications of band-3, together with the neosynthesis of HRP 1, sequestrin and possibly other proteins awaiting clearer identification, represent the products of the biosynthetic program carried out by *P. falciparum* during the intra-erythrocytic part of its life cycle, which includes knob formation and leads to PRBC stickiness. *P. falciparum*-encoded enzymes capable of cleaving cytoskeletal components have recently been described (*Duguercy et al., 1990*).

Three modifications of band-3 have been characterized so far, which are probably responsible for the new arrangement of the intra-membranous particles in PRBC: band-3 modified protein 85 kDa (*Winograd and Sherman, 1989a*), band-3 modified protein 65 kDa (*Crandall and Sherman, 1991*) and band-3 modified protein 240 kDa (*Winograd and Sherman, 1989b*). Band-3 modified protein 85 kDa was noticed by Winograd and Sherman (*1989a*) as the antigen of the monoclonal antibody 4A3, which stains the surface of PRBC in immunofluorescence and immunoelectron microscopy assays and prevents PRBC adhesion to melanoma cells in a test of cytoadherence, in addition to immunoprecipitate a 85 kDa molecule from surface iodinated PRBC. Band-3 modified protein 85 kDa is a knob-associated resident of the outer membrane of PRBC which mediates adhesion to the endothelium, probably by recognition of an unknown cognate ligand on the EC. That it is not a parasite-encoded protein is directly shown by the fact that it does not incorporate radioactive amino acids in labelling experiments and indirectly by the striking similarity between its two-dimensional peptide map with the one of band-3 (*Winograd and Sherman, 1989a; Crandall and Sherman, 1991*). It can be concluded that band-3 modified protein 85 kDa is the result of a proteolytic intervention of *P. falciparum* on band-3, which consists of the small deletion of the amino- and the carboxy-terminal parts of this molecule (*Winograd and Sherman, 1989a; Crandall and Sherman, 1991*). Band-3 modified protein 65 kDa was also noticed as the target antigen of a monoclonal antibody, named 1C4 in this case (*Crandall and Sherman, 1991*). Immunofluorescence techniques show that 1C4 stains the surface of trophozoite- and schizont-PRBC with a granular pattern, and a test of cytoadherence demonstrates that it blocks the adhesion of PRBC to melanoma cells in a dose-dependent manner (*Crandall and Sherman, 1991*). Akin to band-3 modified protein 85 kDa, band-3 modified protein 65 kDa is a component of the outer PRBC membrane, probably knob-associated, which mediates adhesion to the endothelium, likely by direct recognition of a hitherto unknown EC ligand. Data have been collected which suggest that band-3 modified protein 65 kDa is the amino-terminal portion of band-3 specifically truncated by *P. falciparum* enzymatic machinery (*Crandall and Sherman, 1991*). The existence of band-3 modified protein 240 kDa was also proved by Winograd and Sherman (*1987 and 1989b*), who recognized it as the antigen of autoantibodies contained in the sera from patients with malaria. These autoantibodies stain in immunofluorescence assays the surface of trophozoite- and schizont-PRBC, especially those which abundantly express knobs, and immunoprecipitate from their iodinated surface a protein of approximately 240 kDa (*Winograd et al., 1987; Winograd and Sherman, 1989b*). Band-3 modified protein 240 kDa is a knob-associated component of the outer membrane of PRBC. Its direct implication in cytoadherence has not been explored. However, this possibility is favored by the analysis of its two-dimensional peptide map which reveals that this protein is merely the result of the molecular association of units of the 85 kDa modification of band-3 (*Winograd and Sherman, 1989b*).

Endothelium adhesiveness (i): the role of TNF. Compelling evidence implicates the immune system, and TNF in particular, in the pathogenesis of CM (*Grau et al., 1989a*).

Originally, observations were made which suggest that with *P. falciparum*, CM tends to occur in humans who are non-immune toward the parasite yet immunocompetent. Indeed, there is higher frequency of CM: 1) in people from non-endemic rather than endemic areas; 2) within endemic areas, in children more than adults; 3) more frequently in children aged between 1 and 6 years than those under 1 year, who do not yet have a fully mature immune system, and those above 6 years, who are likely to have acquired immunity; 4) in healthy children than malnourished ones (malnutrition is a well known immune deficiency *per se*) (*Marsh and Greenwood, 1986; Phillips and Warrell, 1986; WHO-MAP, 1986; Warrell, 1987*).

Working on the hypothesis that a virgin and effective immune system may mount in some individuals an excessive reaction to *P. falciparum*, including over-production of TNF - the event that proved crucial in experimental CM (see below) - and subsequent CM, TNF serum levels were measured in Malawian children with CM. The results obtained confirmed the starting premise: 1) TNF serum levels are not only higher during malaria than during convalescence, but they are also higher in patients who die than those who survive; 2) TNF serum levels are not only higher in those patients with four or more risk factors than those with less than four, but they are also higher in those with lower (and more ominous) coma scores (1 and 0 according to the Glasgow coma scale); 3) TNF serum levels are higher in those patients who develop post-malaria neurological sequelae than those who completely overcome any neurological complication (*Grau et al., 1989b*).

As part of its pro-inflammatory function, TNF activates the EC (*Mantovani and*

24

Dejana, 1989 and 1992). Among other effects - such as the stimulation of the production of cytokines (e.g., interleukin-1), to amplify the inflammatory reaction; of coagulation and fibrinolysis factors (e.g., thrombospondin, plasminogen activator inhibitor) and platelet activating factor, to promote thrombosis; and of PGI_2 and nitric oxide, to dilate the capillary lumen - TNF induces the EC to upregulate the expression of adhesion molecules on its surface, to favor WBC margination and diapedesis (*Dustin and Springer, 1988; Mantovani and Dejana, 1989; Osborn et al., 1989; Rice and Bevilacqua, 1989; Mantovani and Dejana, 1992*). Among these adhesion molecules, two have been confirmed to be over-expressed on the endothelium of the brain microcirculation during CM and are implicated in cytoadherence of PRBC to the endothelium: CD36 and CD54 (ICAM-1) (Fig. 4) (*Howard and Gilladoga, 1989; Aikawa et al., 1990; Chulay and Ockenhouse, 1990; Hommel, 1990; Ockenhouse et al., 1992b; Sharma, 1991; Sherman et al., 1992*). Data published very recently, moreover, show that VCAM-1 and E-selectin also share similar features (*Ockenhouse et al., 1992b*).

   Endothelium adhesiveness (ii): molecular effectors. Five proteins have been demonstrated to be responsible for the adhesiveness displayed by the endothelium for PRBC: thrombospondin, CD36, ICAM-1 (*Howard and Gilladoga, 1989; Chulay and Ockenhouse, 1990; Hommel, 1990; Sharma, 1991; Sherman et al., 1992*), VCAM-1 and E-selectin (*Ockenhouse et al., 1992b*). Their number seems to be growing, however, and it can not be ruled out that other molecules may play a similar role. It should be noted that, while thrombospondin is secreted by the EC in soluble form and brings the EC and the PRBC into contact by binding to one of two specific receptors, it has on the EC (one of which is CD36 - see below) and to another it has on the PRBC, CD36, ICAM-1, VCAM-1 and E-selectin are membrane proteins on EC surface and directly recognize a respective receptor on the PRBC (Fig. 5).

   Thrombospondin, i.e., thrombin-sensitive protein, is a heavy MW (420 kDa) plasma protein, which is composed of three identical subunits and is secreted by platelets, EC, monocytes/macrophages and fibroblasts (*Lawler and Hynes, 1986; Schwartz, 1989; Hynes, 1991*). Two molecules on EC surface can bind thrombospondin, and one has been recognized as CD36 (*Lawler and Hynes, 1986; Asch et al., 1987; Sun et al., 1989*). Roberts *et al* (*1985*) reported that thrombospondin, adsorbed onto plastic, specifically bound PRBC and that soluble thrombospondin could reverse this phenomenon. It was also observed that thrombospondin helps the adhesion of PRBC to C32 amelanotic melanoma cells (which are CD36-positive) (*Sherwood et al., 1987*). The ligand for thrombospondin on PRBC is yet to be defined. However, evidence indicates that it is a trypsin-sensitive molecule, uniquely located on PRBC surface, in association with knobs (*Sherwood et al., 1987; Nakamura et al., 1992*).

   CD36 (gpIIIb or gpIV) is a 88 kDa membrane protein expressed on the surface of EC, platelets and monocytes/macrophages (*Tandon et al., 1989*). The C32 cells and the U937 myelomonocytic cells used in PRBC cytoadherence assays also express CD36 (*Tandon et al., 1989*). Thrombospondin and type-1 collagen are its natural ligands (*Asch et al., 1987*). Different groups of researchers recognized CD36 as an endothelial molecule which promotes PRBC adhesion. Their interest focused on CD36 since evidence had been provided that thrombospondin binds PRBC (see above). Ockenhouse and Chulay (*1988*) first realized the ability of an anti-CD36 monoclonal antibody (OKM5 - another anti-CD36 monoclonal antibody of common use is OKM8, as above mentioned) to inhibit the binding of PRBC to CD36-positive cells (the U937). They later showed that PRBC adhere to CD36 immobilized on plastic and to cultured EC and to another line of CD36-positive cells (the C32). Interestingly, they also showed that purified and soluble CD36 reversed PRBC adhesion. Barnwell *et al* (*1989*) and Oquendo *et al.* (*1989*) extended and deepened these observations by respectively showing that CD36 can bind directly to PRBC independently of thrombospondin mediation, and that, upon transfection with CD36, not with CD25 used as control, the CD36-negative COS cells acquire the property to bind PRBC. Sequestrin is the ligand for CD36 on PRBC (see above).

   ICAM-1 is a 90 kDa protein expressed on the surface of EC and WBC (*Springer, 1990*). It is an adhesion molecule of the immunoglobulin superfamily (*Staunton et al., 1988*). The integrins CD11a/CD18 (LFA-1) and CD11b/CD18 (CR3) are its natural ligands (*Marlin and Springer, 1987; Diamond et al., 1991*). It has also been observed that rhinoviruses bind it (*Greve et al., 1989*). The finding that PRBC could bind some EC (the human umbilical vein EC) independently of thrombospondin

25

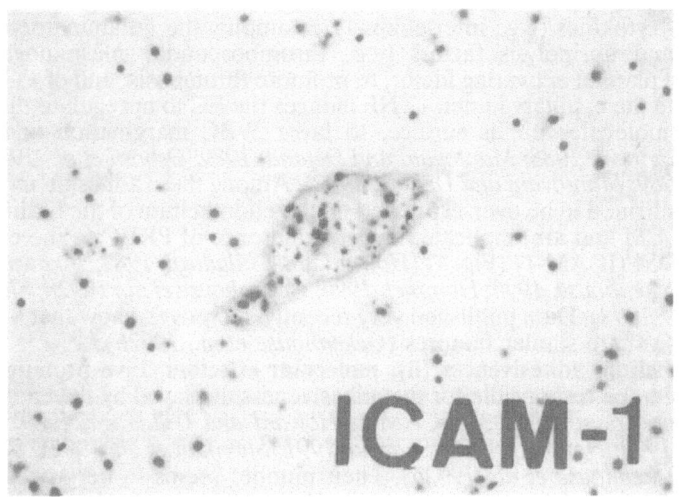

Figure 4: CD54-positive brain microcirculation endothelium in human cerebral malaria. The immunohistochemical staining of a brain post-capillary venule from a patient who died from cerebral malaria by using anti-CD54 (ICAM-1) antibody and peroxidase-conjugated second reagent, shows intense ICAM-1 expression on the endothelium.

and CD36, prompted Berendt *et al* (*1989*) to look for additional molecules on EC capable of binding PRBC. They identified ICAM-1, which mediates PRBC adhesion to HUVEC and, once transfected to the ICAM-1-negative COS cells, renders these capable of binding PRBC. An anti-ICAM-1 monoclonal antibody could reverse PRBC adhesion to either HUVEC or ICAM-1 transfected COS cells (*Berendt et al., 1989*). Ockenhouse *et al.* (*1992a*) and Berendt *et al.* (*1992*) later found that PRBC bind ICAM-1 at a site distinct from LFA-1, CR3 and rhinoviruses but partially overlapping with the site bound by LFA-1. The ligand for ICAM-1 on PRBC remains unknown.

VCAM-1 is a 100 kDa molecule expressed on the surface of activated EC, which belongs to the immunoglobulin superfamily. The integrin VLA-4 is its cognate ligand, which is displayed by monocytes and lymphocytes (*Osborn et al., 1989; Springer, 1990*). E-selectin, also known as ELAM-1, is a 115 kDa protein expressed on surface by activated EC. It is an adhesion molecule of the selectin family, and it possesses an amino-terminal lectin-like domain. It mediates the cytoadherence of WBC to the endothelium by binding the sialyl-Lewis X determinant and probably other carbohydrate extremities on glycosylated receptor molecules (*Bevilacqua et al., 1989; Springer, 1990*). Ockenhouse *et al.* (*1992b*) made the *in vitro* observations that VCAM-1 and E-selectin mediate PRBC cytoadherence to TNF activated HUVEC and that VCAM-1 and E-selectin purified molecules promote PRBC adhesion onto plastic. Telling of the potential role played by these two membrane proteins in CM is the further observation which these researchers made that brain sections from patients who died from CM show microvascular endothelium which expresses on its surface upregulated VCAM-1 and E-selectin, in addition to ICAM-1 and CD36, as already found previously (see above).

CYTOADHERENCE IN MURINE CM

This phenomenon chiefly stems from the stickiness acquired by the mononuclear WBC and the enhanced adhesiveness displayed by the endothelium in the brain microcirculation. It is consequence of the expression of adhesion molecules on both the cellular counterparts involved, which is upregulated by TNF, on its turn over-

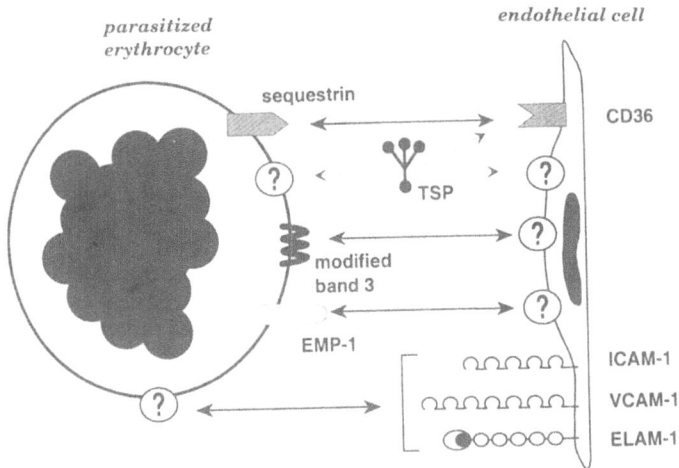

Figure 5: Adhesion molecule interactions mediating cytoadherence of a parasitized red blood cell to the endothelium. TSP: thrombospondin. EMP-1: erythrocyte membrane protein-1. ICAM-1: intercellular cell adhesion molecule-1. VCAM-1: vascular cell adhesion molecule. ELAM-1: endothelial leukocyte adhesion molecule, also known as E-selectin. The CD36 receptor on the parasitized erythrocyte is an identified molecule named sequestrin. The thrombospondin receptors (exception made for CD36), on both the parasitized erythrocyte and the endothelial cell, the receptors for ICAM-1, VCAM-1, and ELAM-1 on the parasitized erythrocyte and the receptors for band-3 modified proteins and EMP-1 on the endothelial cell have not yet been characterized at the molecular level.

produced during the immune response to *P. berghei* ANKA in CM-susceptible strains of mice.

Mononuclear WBC stickiness and endothelium adhesiveness (i): the role of TNF. Extensive evidence for an immunomediated pathogenesis of CM has been collected by studying a murine model of CM (*Grau et al., 1989a*). In this model, animals of susceptible strains (e.g., CBA/Ca) develop lethal CM about a week after infection with *P. berghei* ANKA, a time when parasitemia ranges around 10%, in contrast with animals of resistant strains (e.g., BALB/c) which do not face any cerebral complication, show progressively increasing parasitemia and die of anaemia during the fourth week of infection, when parasitemia can be as high as 70% (*Grau et al., 1989a*). The phenotypic hallmark of this model of CM is a deep comatous state of sudden onset and rapid evolution, and the histological one is the presence of cellular plugs in the brain capillaries and post-capillary venules, which in turn show necrotic walls and ring hemorrhages all around (*Grau et al., 1989a*). Human and murine CM share common features of both neurological and histological nature, with exception made as to the cellular type forming the vessel plug, being chiefly the PRBC in man and the mononuclear WBC in the mouse (*MacPherson et al., 1985; Grau et al., 1989a*).

That the immune system plays a central role in the development of CM became evident when it was observed that the athymic nude variant of CM-susceptible mice does not develop CM upon *P. berghei* ANKA infection (*Finley et al., 1982*). This prompted a number of studies to follow, reviewed by Grau *et al.* (*1991a*), which illustrated the contribution of cytokines, first of all TNF, to the development of CM (Fig. 6). The identification of the central role of TNF in CM rests on the following facts: 1) TNF serum levels dramatically raise during the few hours preceding CM development, and this seems to be a highly specific event, which takes place only in susceptible mice infected with *P. berghei* ANKA and not in resistant mice or in susceptible mice infected by a *Plasmodium* unable to cause CM, such as *P.yoelii*; 2) treatment with anti-TNF antibody prevents CM onset in susceptible mice; 3) administration of recombinant TNF confers susceptibility to CM to CM-resistant mice

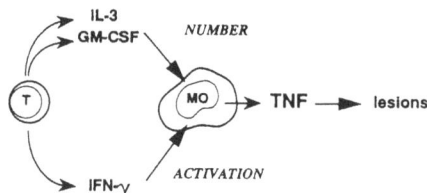

Figure 6: Cytokine cascade leading to TNF over-production in murine CM. For TNF to be produced in excessive amounts, monocytes/macrophages need to be in high number (the enlargement of monocyte/macrophage pool requires the action of interleukin-3 and granulocyte/macrophage-colonystimulating factor) and to be activated by interferon-gamma (*reproduced from Grau et al., 1989a*).

Grau et al., 1987). Interestingly, plugging of cerebral vessels with mononuclear WBC can be observed in the absence of plasmodial infection in CM-susceptible mice given high doses of recombinant TNF (*Grau et al., 1989a*).

The link in the pathogenetic sequence of events leading to murine CM between TNF over-production and cytoadherence of mononuclear WBC to the brain microcirculation endothelium is represented by over-expression of adhesion molecules on both the cellular components involved. As mentioned above in regard to the EC, this phenomenon is induced by TNF as an effect of its pro-inflammatory action, and it has been documented by immunohistochemistry studies in murine CM (*Grau et al., 1991b*) (Fig. 7). TNF can upregulate the expression of adhesion molecules, not only on the EC, but also on circulating monocytes and lymphocytes (*Trinchieri, 1992*).

Mononuclear WBC stickiness and endothelium adhesiveness (ii): molecular effectors. Cytoadherence between WBC and endothelium is assured by different adhesion molecules, which have recently been reviewed elsewhere (*Osborn, 1990; Zimmerman et al., 1992; Shimizu et al., 1992; Makgoba et al., 1992*). Each of these molecules has another one as a cognate ligand on the cellular counterpart, and it is via their reciprocal, specific binding that WBC and EC eventually come into contact. Adhesion molecules which are expressed by mononuclear WBC and which bind ligands on the endothelium are: the integrins LFA-1 and CR3, which recognize ICAM-1 as their ligand on the EC (see above), and VLA4, which recognizes VCAM-1 as its EC ligand (*Osborn et al., 1989*); the L-selectin, which is a lectin and can therefore bind the extremities of carbohydrate chains, but for which no molecule has been identified as its specific ligand on the EC (*Brandley et al., 1990*); and CD44, which binds hyaluronate, but no specific ligand on the EC hitherto characterized (*Haynes et al., 1989*).

The adhesion molecules responsible for the cytoadherence of the mononuclear WBC to the endothelium in the brain microcirculation in murine CM have not yet been identified. Although clear data incriminate the LFA-1/ICAM-1 interaction in the pathogenesis of murine CM, this can not be exactly held responsible for the cytoadherence of the mononuclear WBC to the endothelium as it is illustrated below.

Falanga and Butcher (*1991*) and Grau *et al.* (*1991b*) tried to prevent murine CM by the injection of monoclonal antibodies directed against different mononuclear WBC adhesion molecules, namely, LFA-1, CR3, L-selectin, CD49d/CD29 (VLA4), CD44 and ICAM-1; out of these reagents only the anti-LFA-1 antibodies (FD445.1, used by Falanga and Butcher and H35.89.9, used by Grau and colleagues) prevented CM, and they did it very effectively. By contrast, the anti-ICAM-1 antibody caused hemoptysis and accelerated mouse death, while it is innocuous in normal mice (*Grau et al., 1991b*). It is important to note that the anti-LFA-1 antibodies, like the other antibodies, were given on the sixth day of infection, a time that just precedes CM onset (see above), by which parasite development and the related immune response (TNF over-production included) had taken place as expected (*Falanga and Butcher, 1991; Grau et al., 1991b*). Grau *et al.* (*1991b*) also tried to cure CM by injecting the anti-LFA-1 antibody within twelve hours after the appearance of its signs: the antibody

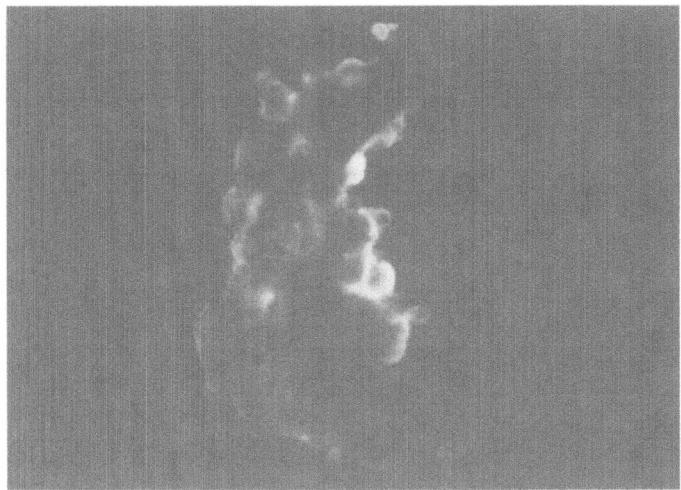

Figure 7: CD54-positive brain microcirculation endothelium in murine cerebral malaria. The immunohistochemical staining of a brain post-capillary venule from a mouse suffering from cerebral malaria by using anti-CD54 (ICAM-1) antibody and fluorescein-conjugated second reagent shows intense ICAM-1 expression on the endothelium and also on cellular components of the vessel plug.

reverted the condition very effectively. However, surprisingly, in contrast to the results at phenotypic level and to the achievement of full survival of the treated mice, anti-LFA-1 administration did not result in the abrogation of the cytoadherence of mononuclear WBC to the endothelium and the ensuing vessel plug formation in the brain capillary and post-capillary venules. Indeed, the cellular plugs were still visible in the successfully treated mice in the same way they were in the animals affected by CM (*Falanga and Butcher, 1991; Grau et al., 1991b*).

VESSEL PLUG FORMATION IN HUMAN AND MURINE CM: COMMON MECHANISMS

The pathogenetic sequence of events that leads to cytoadherence and vessel plug formation in human and murine CM presents important similarities. Attention should not be distracted from these by differences in the cellular components of the vessel plug seen in the two conditions (Fig. 8).

The development of both human and murine CM follows: 1) a potent immune response of the mammalian host raised by the infecting *Plasmodium*, which includes a cytokine cascade with TNF over-production; 2) the upregulation of adhesion molecules on the endothelium of brain capillaries and post-capillary venules, which is induced by TNF; 3) cytoadherence of circulating sticky cells to the highly adhesive activated endothelium, i.e., chiefly PRBC in human CM, which are equipped to recognize CD36, ICAM-1, VCAM-1 and ELAM-1 on the EC and also to exploit thrombospondin to adhere to it, and mononuclear WBC in murine CM, which express upregulated LFA-1 and probably other molecules sticky for the endothelium (see above). Thus, PRBC are by far the main component of the vessel plug in human CM since they are by far the most represented circulating cell type capable of binding adhesion molecules on the endothelium. Indeed, WBC, although endowed with similar capability, are less represented in the vessel plug since they are outnumbered by PRBC in the bloodstream. It is interesting to emphasize in this context that the presence of cells of the monocyte/macrophage lineage can be invariably documented

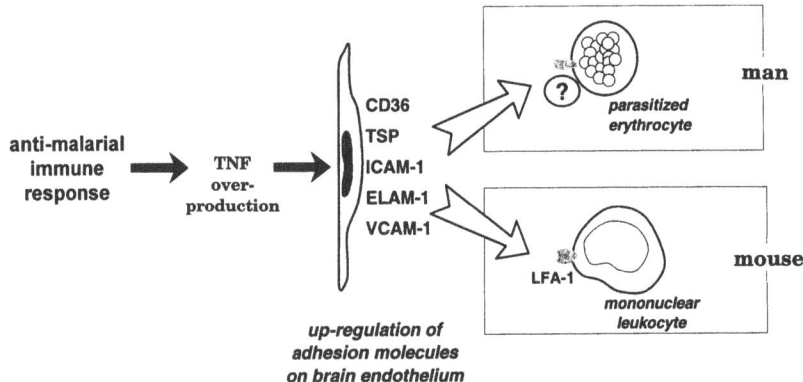

Figure 8: Common mechanisms in vessel plug formation in human and murine CM.

in brain post-capillary venules in human CM, at least when it is looked for, by using an anti-CD68 monoclonal antibody in immunohistochemical techniques (*J. Porta, unpublished results*). Platelets, which are LFA-1-negative in humans, have no stickiness for a non-activated, non-thrombogenic endothelium, which is probably already largely bound by PRBC at the time it acquires in *P. falciparum* infection, upon inflammatory stimuli, platelet adhering properties. Likewise, mononuclear WBC are the main component of the vessel plug in murine CM, because no other circulating cell types express surface molecules to compete with them for binding to the adhesion molecules displayed by the endothelium.

ANTI-ADHESION INTERVENTION IN CM

This can be envisaged in an either preventive or therapeutic setting. Prevention is the only strategy to pursue in the fight against malaria, and it is now clear that only active immunization holds hopes of feasibility. However, despite conspicuous research efforts, the preparation of an effective vaccine has failed thus far (*Miller et al., 1986*). The realization of the appearance of novel sticky proteins on the PRBC and of their relevance to the pathogenesis of CM opens a new prospective of preventive immunointervention against malaric pathology (see above). In fact, these proteins, or their derivatives, are candidates for a vaccine which, although it cannot aim at eradicating the plasmodial infection, could prevent the development of CM by potentiating the immune response against PRBC sticky parts and deterring cytoadherence as a consequence. This would be an important achievement, given the contribution of the cerebral complication to the toll of deaths that malaria continuously exacts. Unfortunately, the immunological diversity of these novel proteins observed among different field isolates (*Sherman et al., 1992*) and data recently appearing in the literature, which shows that these proteins are among the antigens that *P. falciparum* subjects to rapid variation (*Roberts et al., 1992*), herald difficulties also concerning the preparation of such an "anti-CM" vaccine.

The armamentarium currently available for the therapy of CM, which consists of anti-*Plasmodium* chemotherapies, needs potentiation, given the remarkable number of cases of failure. A new option is represented by treatment with anti-TNF antibodies and a monoclonal reagent is now under phase II trial in the Gambia (*D. Kwiatowski et al., submitted for publication*). PRBC stickiness is a potential target of therapeutic intervention against CM: if it is neutralized, the cytoadherence that generates CM can not take place. The use can therefore be envisaged of reagents aimed at blocking PRBC sticky molecules, such as antibodies or binding peptides or even endothelium adhesion molecules in soluble form. The employment of the latter has the attraction of dodging the problem of parasite variation. Indeed, while the novel sticky molecules expressed on PRBC surface could, by changing, escape the surveillance of antibodies and binding peptides, they cannot lose affinity for at least a region on the adhesion molecules displayed by the human endothelium: if they did, they would stop being sticky molecules by definition. Research in this direction deserves to be encouraged.

Footnote

Throughout the paper we refer to stickiness as the property of circulating cells to adhere to the endothelium and to adhesiveness as the property of the endothelium to bind circulating cells.

REFERENCES

Ahlborg, N., Berzins, K., Perlmann, P., 1991, Definition of the epitope recognized by the Plasmodium falciparum-reactive human monoclonal antibody 33G2. *Mol. Biochem. Parasitol.* 46:89.
Aikawa, M., 1988, Human cerebral malaria. *Am. J. Trop. Med. Hyg.* 39:3.
Aikawa, M., Atkinson, C.T., 1990, Immunoelectron microscopy of parasites. *Adv. Immunol.* 29:151.
Aikawa, M., Iseki, M., Barnwell, J.W., Taylor, D., Oo, M.M., Howard, R.J., 1990, The pathology of human cerebral malaria. *Am. J. Trop. Med. Hyg.* 43:S30.
Aikawa, M., Udeinya, I.J., Rabbege, J., Dayan, M., Leech, J.H., Howard, R.J., Miller, L.H., 1985, Structural alterations of the membrane of erythrocytes infected with Plasmodium falciparum. *J. Protozol.* 32:424.
Allred, D.R., Gruenberg, J.E., Sherman I.W., 1986, Dynamic rearrangements of erythrocyte membrane internal architecture induced by infection with Plasmodium falciparum. *J. Cell Sci.* 81:1.
Asche, A.S., Barnwell, J., Silverstein, R.L., Nachman, R.L., 1987, Isolation of the thrombospondin membrane receptor. *J. Clin. Invest.* 79:1054.
Barnwell, J.W., Asch, A.S., Nachman, R.L., Yamaya, M., Aikawa, M., Ingravallo, P., 1989, A human 88-kDa membrane glycoprotein (CD36) functions *in vitro* as a receptor for a cytoadherence ligand on Plasmodium falciparum-infected erythrocytes. *J. Clin. Invest.*, 84:765.
Berendt, A.R., McDowall, A., Graig, A.G., Bates, P.A., Sternberg, M.J.E., Marsh, K., Newbold, C.I., Hogg, N., 1989, The binding site on ICAM-1 for Plasmodium falciparum-infected erythrocytes overlaps, but is distinct from, the LFA-1 binding site. *Cell*, 68:71.
Berendt, A.R., Simmons, D.L., Tansey, J., Newbold, C.I., Marsh, K., 1992, Intercellular adhesion mole cule-1 is an endothelial cell adhesion receptor for Plasmodium falciparum. *Nature* 341:57.
Bevilacqua, M.P., Stengelin, S., Gimbrone Jr., M.A., Seed, B., 1989, Endothelial leukocyte adhesion molecule 1: an inducible receptor for neutrophils related to complement regulatory proteins and lectins. *Science* 243:1160.
Brandley, B.K., Swiedler, S.J., Robbins, P.W., 1990, Carbohydrate ligands of the LEC cell adhesion molecules. *Cell*, 63:861.
Chulay, J.D., Ockenhouse, C.F., 1990, Host receptors for malaria-infected erythrocytes. *Am. J. Trop. Med. Hyg.*, 43:S6.
Crandall, I., Sherman, I.W., 1991, Plasmodium falciparum (human malaria)-induced modifications in human erythrocyte band-3 protein. *Parasitology* 102:335.
Culvenor, J.G., Day, K.P., Anders, R.F., 1991,. Plasmodium falciparum ring-infected erythrocyte surface antigen is released from merozoite dense granules after erythrocyte invasion. *Infect. Immun.*, 59:1183.
Diamond, M.S., Staunton, D.E., Marlin, S.D., Springer, T.A., 1991, Binding of the integrin Mac-1 (CD11b/CD18) to the third immunoglobulin-like domain of ICAM-1 (CD54) and its regulation by glycosylation. *Cell* 65:961.
Duguercy, A., Hommel, M., Schrevel, J., 1990, Purification and characterization of 37 kDa proteases from P. falciparum and P. berghei which cleaves erythrocyte cytoskeletal components. *Mol. Biochem. Parasitol.* 38:233.
Dustin, M.L., Springer, T.A., 1988, Lymphocyte function-associated antigen-1 (LFA-1) interaction with intercellular adhesion molecule-1 (ICAM-1) is one of at least three mechanisms for lymphocyte adhesion to cultured endothelial cells. *J. Cell Biol.* 107:321.
Falanga, P.B., Butcher, E.C., 1991, Late treatment with anti-LFA-1 (CD11a) antibody prevents cerebral malaria in a mouse model. Eur. *J. Immunol.* 21:2259.

Finley, R.W., Mackey, L.J., Lambert, P.-H., 1982, Virulent P.berghei malaria: prolonged survival and decreased cerebral pathology in T-cell deficient nude mice. *J. Immunol.* 129:2213.

Grau, G.E., Bieler, G., Pointaire, P., De Kossodo, S., Tacchini-Cottier F., Piguet, P.-F., Vassalli, P., Lambert, P.-H., 1990, Significance of cytokine production and adhesion molecules in malarial immunopathology. *Immunology Letters* 25:189.

Grau, G.E., Fajardo, L.F., Piguet, P.-F., Allet, B., Lambert, P.-H., 1987, Tumor necrosis factor/cachectin as an essential mediator in murine cerebral malaria. *Science* 237:1210.

Grau, G.E., Piguet, P.-F., Vassalli, P., Lambert, P.-H., 1989a, Tumor necrosis factor and other cytokines in cerebral malaria: experimental and clinical data. *Immunol. Reviews* 112:49.

Grau, G.E., Pointaire, P., Piguet, P.-F., Vesin, C., Rosen, H., Stamenkovic, I., Takei, F., Vassalli, P., 1991, Late administration of monoclonal antibody to leukocyte function-antigen 1 abrogates incipient murine cerebral malaria. *Eur. J. Immunol.* 21:2265.

Grau, G.E., Taylor, T.E., Molyneux, M.E., Wirima, J.J., Vassalli, P., Hommel, M., Lambert, P.-H., 1989b, Tumor necrosis factor and disease severity in children with falciparum malaria. *N. Engl. J. Med.* 320:1586.

Greenwood, B.M., Bradley, A.K., Greenwood, A.M., Byass, P., Jammen, K., Marsh, K., Tulloch, S., Oldfiled, F.S., Hayes, R., 1987, Mortality and morbidity from malaria among children in a rural area of The Gambia, West Africa. *Trans. R. Soc. Trop. Med. Hyg.* 91:478.

Greve, J.M., Davis, G., Meyer, A.M., Forte, C.P., Yost, S.C., Marlor, C.W., Kamarck, M.E., McClelland, A., 1989, The major human rhinovirus receptor is ICAM-1. *Cell* 56:839.

Gruenberg, J., Allred, P.R., Sherman, I.W., 1983, A scanning electron microscope analysis of the protrusions (knobs) present on the surface of Plasmodium falciparum-infected erythrocytes. *J. Cell Biol.* 97:795.

Haynes, B.F., Telen, M.J., Hale, L.P., Denning, S.M., 1989, CD44 - a molecule involved in leukocyte adherence and T-cell activation. *Immunol. Today* 10:423.

Hommel, M., 1990, Cytoadherence of malaria-infected erythrocytes. *Blood Cells* 16:605.

Howard, R.J., Barnwell, J.W., Rock, E.P., Jeequaye, J., Ofori-Adjei D., Maloy, W.L., Lyon, J.A., Saul, A., 1988, Two approximately 300 kilodalton Plasmodium falciparum proteins at the surface membrane of infected erythrocytes. *Mol. Biochem. Parasitol* 27:207.

Howard, R.J., Gilladoga, A.D., 1989, Molecular studies related to the pathogenesis of cerebral malaria. *Blood* 74:2603.

Howard, R.J., Handunetti, S., Hasler, T., Gilladoga, A., de Aguiar, J., Pasloske, B., Morehaed, D., Albrect, G., van Schravendijik, M., 1990, Surface molecules on Plasmodium falciparum-infected erythrocytes involved in adherence. *Am. J. Trop. Med. Hyg.* 43:S15.

Howard, R.J., Lyon, J.A., Uni, S., Saul, A.J., Aley, S.B., Klotz, F., Panton, L.J., Sherwood, J.A., Marsh, K., Aikawa, M., Rock, E.P., 1987, Transport of an M_r = 300 000 Plasmodium falciparum protein (PfEMP2) from the intraerythrocytic asexual parasite to the cytoplasmic face of the host cell membrane. *J. Cell. Biol.* 104:1269.

Howard, R.J., Uni, S., Aikawa, M., Aley, S.B., Leech, J.H., Lew, A.M., Wellems, T.E., Marsh, K., Rener, J., Taylor, D.W., 1986, Secretion of a malaria histidin-rich protein (Pf HRP II) from Plasmodium falciparum-infected erythrocytes. *J. Cell Biol.* 103:1269.

Hynes, R.O., 1991, The complexity of platelet adhesion to extracellular matrices. *Thromb. Hemostasis* 66:40.

Kilejian, A., Jensen, J.B., 1977, A histidine-rich protein from P.falciparum and its interaction with membranes. *Bull. Wld. Hlth. Org.* 55:191.

Langreth, S.G., Jensen, J.B., Reese, R.T., Trager, W., 1978, Fine structure of human malaria *in vivo*. *J. Protozol.* 25:443.

Lawler, J., Hynes, R.O., 1986, The structure of human thrombospondin, an adhesive glycoprotein with multiple calcium-binding sites and homologies with several different proteins. *J. Cell Biol.* 103:1635.

Leech, J.H., Barnwell, J.W., Miller, L.H., Howard, R.J., 1984, Identification of a strain-specific malarial antigen exposed on the surface of Plasmodium falciparum-infected erythrocytes. *J. Exp. Med.* 159:1567.

Low, P.S., 1986, Structure and function of the cytoplasmic domain of band-3: center of erythrocyte membrane-peripheral protein interactions. *Biochim. Biophys. Acta* 864:145.

MacPherson, G.G., Warrell, M.J., White, N.J., Looareesuwan, S., Warrell, D.A., 1985, Human cerebral malaria. A quantitative ultrastructural analysis of parasitized erythrocyte sequestration. *Am. J. Pathol.* 119:385.

Maguire, P.A., Sherman, I.W., 1990, Phospholipid composition, cholesterol content and cholesterol exchange in Plasmodium falciparum-infected red cells. *Mol. Biochem. Parasitol.* 38:105.

Makgoba, M.W., Bernard, A., Sanders, M.E., 1992, Cell adhesion/signalling: biology and clinical applications. *Eur. J. Clin. Invest.* 22:443.

Mantovani, A., Dejana, E., 1989, Cytokines as communication signals between leukocytes and endothelial cells. *Immunol. Today,* 10:370.

Mantovani, A., Dejana, E., 1992, Functional responses elicited in endothelial cells by cytokines. In: Kunkel, S.L. and Remick, D.G. (eds); Cytokines in health and disease. *Dekker, New York* 297-307.

Marlin, S.D., Springer, T.A., 1987, Purified intercellular adhesion molecule-1 (ICAM-1) is a ligand for lymphocyte function-associated antigen-1 (LFA-1). *Cell* 51:813.

Marsh, K., Greenwood, B.M., 1986, Immunopathology of malaria. *Clin. Trop. Med. Communicable Dis.* 1:91.

Miller, L.H., Howard, R.J., Carter, R., Good, M.F., Nussenzweig, V., Nussenzweig, R., 1986, Research toward malaria vaccine. *Science* 234:1349.

Nakamura, K.-I., Hasler, T., Morehead, K., Howard, R.J., Aikawa, M., 1992, Plasmodium falciparum-infected erythrocyte receptor(s) for CD36 and thrombospondin are restricted to knobs on the erythrocyte surface. *J. Histochem. Cytochem.* 40:1419.

Nash, G.B., O'Brien, E., Gordon-Smith, E.C., Dormandy, J.A., 1989, Abnormalities in the mechanical properties of red blood cells caused by Plasmodium falciparum. *Blood* 74:855.

Ockenhouse, C.F., Betageri, R., Springer, T.A., Staunton, D.E., 1992a, Plasmodium falciparum-infected erythrocytes bind ICAM-1 at a site distinct from LFA-1, Mac-1, and human rhinovirus. *Cell* 68:63.

Ockenhouse, C.F., Chulay, J.D., 1988, Plasmodium falciparum sequestration: OKM5 antigen (CD36) mediates cytoadherence of parasitized erythrocytes to a myelomonocitic cell line. *J. Infect. Dis.* 157:584.

Ockenhouse, C.F., Klotz, F.W., Tandon, N.N., Jamieson, G.A., 1991, Sequestrin, a CD36 recognition protein on Plasmodium falciparum-infected erythrocytes identified by anti-idiotype antibodies. *Proc. Natl. Acad. Sci. U.S.A.,* 88:3175.

Ockenhouse, C.F., Tegoshi, T., Maeno, Y., Benjamin, C., Ho, M., Kan, K.E., Thway, Y., Win, K., Aikawa, M., Lobb, R.R., 1992b, Human vascular endothelial cell adhesion receptors for Plasmodium falciparum-infected erythrocytes: roles for endothelial leukocyte adhesion molecule 1 and vascular cell adhesion molecule 1. *J. Exp. Med.* 176:1183.

Oo, M.M., Aikawa, M., Than, T., Aye, T.M., Myint, P.T., Igarashi, I., Schoene, W.C., 1987, Human cerebral malaria: a pathological study. *J. Neuropathol. Exp. Neurol.* 46:223.

Osborn, L., Hession, C., Tizard, R., Vassallo, C., Luhowskyj, S., Chi-Rosso, G., Lobb, R., 1989, Direct expression cloning of vascular cell adhesion molecule 1, a vascular cytokine-induced endothelial protein that binds to lymphocytes. *Cell* 59:1203.

Osborn, L., 1990, Leukocyte adhesion to endothelium in inflammation. *Cell* 62:3.

Oquendo, P., Hundt, E., Lawler, J., Seed, B., 1989, CD36 directly mediates cytoadherence of Plasmodium falciparum parasitized erythrocytes. *Cell* 58:95.

Petersen, C., Nelson, R., Leech, J., Jensen, J., Wollish, W., Scherf, A., 1990, The gene product of the Plasmodium falciparum 11.1 locus is a protein larger than one megadalton. *Mol. Biochem. Parasitol.* 42:189.

Phillips, R.E., Warrell, D.A., 1986, The pathophysiology of severe falciparum malaria. *Parasitol. Today* 2:271.

Rice, G.E., Bevilacqua, M.P., 1989, An inducible endothelial cell surface glycoprotein mediates melanoma adhesion. *Science* 246:1303.

Roberts, D.D., Sherwood, J.A., Spitalnik, S.L., Panton, L.J., Howard, R.J., Dixit, V.M., Frazier, W.A., Miller, L.H., Ginsburg, V., 1985, Thrombospondin binds falciparum malaria parasitized erythrocytes and may mediate cytoadherence. *Nature,* 318:64.

Roberts D.J., Craig, A.J., Berendt, A.R., Pinches, R., Nash, G., Marsh, K., Newbold C.I., 1992, Rapid switching to multiple antigenic and adhesive phenotypes in malaria. *Nature* 357:689.

Roman, G.C., 1991, Cerebral malaria: the unsolved riddle. *J. Neurol. Sci.* 101:1.

Ruoslahti, E., Giancotti, F.G., 1989, Integrins and tumor cell dissemination. *Cancer Cells* 1:119.

Schwartz, B.S., 1989, Monocyte synthesis of thrombospondin: the role of platelets. *J Biol. Chem.* 264:7512.

Sharma, Y.D., 1991, Knobs, knob proteins and cytoadherence in Plasmodium falciparum malaria. *Int. J. Biochem.* 23:775.

Sherman, I.W., Crandall I., Smith, H., 1992, Membrane proteins involved in the adherence of Plasmodium falciparum-infected erythrocytes to the endothelium. *Biol. Cell.,* 74:161.

Sherwood, J.A., Roberts, D.D., Marsh, K., Harvey, E.B., Spitalnik, S.L., Miller, L.H., Howard, R.J., 1987, Thrombospondin binding by parasitized erythrocyte isolates in falciparum malaria. *Am. J. Trop. Med. Hyg.*, 36:228.

Shimizu, Y., Newman, W., Tanaka, Y., Shaw, S., 1992, Lymphocyte interactions with endothelial cells. *Immunol. Today* 13:106.

Springer, T.A., 1990, Adhesion receptors of the immune system. *Nature*, 346:425.

Staunton, D.E., Marlin, S.D., Stratowa, C., Dustin, M.L., Springer, T.A., 1988, Primary structure of ICAM-1 demonstrates interaction between members of the immunoglobulin and integrin supergene families. *Cell* 52:925.

Sun, X., Mosher, D.F., Rapraeger, A., 1989, Heparin sulphate-mediated binding of epithelial cell surface proteoglycan to thrombospondin. *J. Biol. Chem.* 264:2885.

Talle, M.A., Rao, P.E., Westberg, E., Allegar, N., Makowski, M., Mittler, R.S., Goldstein, G., 1983, Patterns of antigenic expression on human monocytes as defined by monoclonal antibodies. *Cell. Immunol.* 78:83.

Tandon, N.N., Lipsky, R.H., Burgess, W.H., Jamieson, G.A., 1989, Isolation and characterization of platelet glycoprotein IV (CD36). *J. Biol. Chem.*, 264:7570.

Taylor, D.2W., Parra, M., Chapman, G.B., Stearns, M.E., Rener, J., Aikawa, M., Uni, S., Aley, S.B., Panton, L.J., Howard, R.J., 1987, Localization of Plasmodium falciparum histidine rich protein in the erythrocyte skeleton under knobs. *Mol. Biochem. Parasitol.* 25:165.

Trager, W., Rudzinska, M.A., Bradbury, P.C., 1966, The fine structure of Plasmodium falciparum and its host erythrocyte in natural malarial infections in man. *Bull. Wld. Hlth. Org.* 35:883.

Trinchieri, G., 1992, Effects of TNF and lymphotoxin on the hematopoietic system. In: Aggarwal, B.B. and Vilcek, J. (eds); *Tumor necrosis factor*. Dekker, New York, 289-313.

Udeinya, I.J., Miller, L.H., McGregor, I.A., Jensen, J.B., 1983, Plasmodium falciparum strain-specific antibody blocks binding of infected erythrocytes to amelanotic melanoma cells. *Nature* 303:429.

Udomsangpetch, R., Aikawa, M., Berzins, K., Wahlgren, M., Perlmann, P., 1989a, Cytoadherence of knobless Plasmodium falciparum-infected erythrocytes and its inhibition by a human monoclonal antibody. *Nature* 338:763.

Udomsangpetch, R., Wahlin, B., Carlson, J., Berzins, K., Torii, M., Aikawa, M., Perlmann, P., Wahlgren, M., 1989b, Plasmodium falciparum-infected erythrocytes form spontaneous erythrocyte rosettes. *J. Exp. Med.* 169:1835.

Warrell, D.A., 1987, The pathophysiology of severe falciparum malaria in man. *Parasitology* 94:S53.

Wernsdorfer, W.H., 1980, The importance of malaria in the world. In: Kreier, J.P. (ed); Malaria. Academic Press, London, 1:93.

White, N.J., 1986, Malaria physiopathology. *Clin. Trop. Med. Communicable Dis.* 1:55.

Winograd, E., Greenan, J.R.T., Sherman, I.W., 1987, Expression of senescent antigen on erythrocytes infected with a knobby variant of the human malaria parasite Plasmodium falciparum. *Proc. Natl. Acad. Sci. U.S.A.* 84:1931.

Winograd, E., Sherman, I.W., 1989a, Characterization of a modified red cell membrane protein expressed on erythrocytes infected with the human malaria parasite Plasmodium falciparum: possible role as a cytoadherent mediating protein. *J. Cell Biol.* 108:23.

Winograd, E., Sherman, I.W., 1989b, Naturally occurring anti-band-3 autoantibodies recognize a high molecular weight protein on the surface of Plasmodium falciparum-infected erythrocytes. *Biochem. Biophys. Res. Comm.* 160:1357.

World Health Organization Malaria Action Program, 1986, Severe and complicated malaria. *Trans. R. Soc. Trop. Med. Hyg.* 80:S1.

Zimmerman, G.A., Prescott, S.M., McIntyre, T.M., 1992, Endothelial cell interactions with granulocytes: tethering and signaling molecules. *Immunol. Today* 13:93

ENDOTHELIAL-TUMOR CELL INTERACTIONS *IN VITRO* AND *IN VIVO*

*R. Giavazzi, *A. Garofalo, **I. Martin-Padura
***A.J.H. Gearing and **E. Dejana

*Mario Negri Institute for Pharmacological Research
24100 Bergamo
Italy

**Mario Negri Institute for Pharmacological Research
20157 Milano
Italy

***British Biotechnology Limited
Cowley, Oxford
United Kingdom

INTRODUCTION

The development of a viable metastasis is a complex multi-step process (*Poste, G. and Fidler, I.J., 1980*). A metastasizing cell must first detach from the primary tumor, degrade and invade the surrounding tissue before hematogenous and lymphatic dissemination can occur. Neoplastic cells are released into the circulation in large numbers, but only a few will establish a metastasis; many cells will die because they are unable to survive the hemodynamic pressure and host defense mechanisms. The arrest in the capillary bed of secondary organs is the next step, followed by extravasation into the organ parenchyma and establishment as micro-metastases. Once growth and neovascularization have occurred, metastatic lesions can produce further metastases. A major mechanism of tumor cell extravasation consists of the rapid attachment to the endothelium, followed by endothelial cell retraction, migration of tumor cells to the subendothelial matrix with the dissolution of the basement membrane (*Crissman et al., 1988; Kramer and Nicolson, 1979; Lapis et al., 1988; Liotta, 1986*). Disruption of the endothelial layer (*Kawaguchi and Nakamura, 1986*) during extravasation has also been described. The initial endothelial cell recognition precedes the extravasation of circulating tumor cells (*Kawaguchi and Nakamura, 1986; Kramer and Nicolson, 1979*). The location of the first capillary bed encountered only partially explains the site of metastasis formation, and tumor types that selectively disseminate to certain organs have been found in experimental and human systems (*Fidler, 1990; Nicolson, 1988; Sugarbaker, 1979*). Preferential adhesion/binding of metastatic cells to endothelial cells from target organs has been described, suggesting that endothelial-tumor cell interactions play a role in metastatic homing to specific sites (*Auerbach et al., 1987; Belloni and Tressler, 1989*). Cell surface adhesion structures, including lectins, proteoglycans and integrins, have been proposed as mediators of tumor-endothelial cell interaction, but their specific roles in metastasis formation still need to be clarified (*Belloni and Tressler, 1989; Raz and Lotan, 1987; Rouslathi and Giancotti, 1989*).

Vascular changes can easily influence the interaction of tumor cells with endothelium and thus the whole metastatic process (*Belloni and Tressler, 1989; Weiss et al., 1989*). Thrombotic events mediate the arrest of tumor cells; the release of chemotactic factors from the vascular wall itself can induce tumor cell motility; extracellular matrix components may modulate endothelial cell recognition structures responsible for tumor interactions. Increased tumor cell localization and metastasis formation can be induced by several factors, including some conventional antineoplastic treatments that can damage the vascular wall causing endothelial cell retraction, resulting in increased attachment of tumor cells to the exposed matrix (*Orr et al., 1986; Van Den Brenk, et al., 1973*). Inflammation may also cause changes in vascular phenotype and tumor secondary localization has often been associated with inflammatory stimuli (*Levine and Saltzman, 1990; Murphy et al., 1988; Sugarbaker, 1979*). Here we will address the role of inflammatory cytokines in the metastatic process, focusing on the possible mechanisms that regulate tumor-endothelial cell interaction under inflammatory stimuli. Other aspects of the interaction of malignant cells with the vasculature, such as the hemostatic-coagulation response associated with tumor cell arrest, and angiogenesis leading to tumor vascularization, are covered in excellent reviews (*Folkman, 1986; Weiss et al., 1989*).

CYTOKINES AFFECT TUMOR CELL ADHESION AND METASTASIS

The availability of human endothelial cell cultures has provided a useful tool for investigating mechanisms that influence tumor-endothelial cell interactions. Activation of cultured human umbilical endothelial cells by inflammatory mediators, such as interleukin-1 (IL-1) and tumor necrosis factor-α (TNF), increases the adhesion of human tumor cells of different histological type and origin (*Dejana et al., 1988; Lauri et al., 1990; Rice et al., 1988*). Augmentation of adhesion was observed for human melanoma, osteosarcoma and colon, renal, lung, mammary carcinoma lines. The relevance of these findings on metastasis formation has been studied in nude mice receiving IL-1. IL-1 (IL-1α and IL-1β), given to mice before intravenous injection of radio-labelled tumor cells, augmented their retention in the lung at 4-24h (*Giavazzi et al., 1990; Lauri et al., 1990*). At autopsy, eight weeks later, mice given IL-1 developed more lung colonies (experimental metastases) than control mice (*Bani et al., 1991; Giavazzi et al., 1990*). The effect of IL-1 on metastasis formation was observed when IL-1 was given 1 to 4 hours before tumor cells or shortly after; no effect on the growth of established metastases was observed (*Giavazzi et al., 1990*). This is in contrast to other reports describing antitumor and antimetastatic effect by IL-1 (*Belardelli et al., 1989; Nakamura et al., 1986*) though different tumor lines and experimental conditions were described in those experiments. Metastasis augmentation was observed with murine (injected in syngeneic mice) and human (injected in nude mice) tumor lines of different histological origin and metastatic nature (*Bani et al., 1991; Giavazzi et al., 1990*) but only with tumors that are prone to metastasize. IL-1 did not induce metastases by nonmetastatic tumors and did not change their pattern of homing. The effect of IL-1 treatment on metastasis formation in syngeneic mice bearing a primary tumor (spontaneous metastases) showed an augmentation of metastasis to the lung in mice with the B16 melanoma (metastatic to the lung) and to the liver in mice with the M5076 reticulum cell sarcoma (metastatic to the liver) (*Bani et al., 1991*). These findings suggest the increase in metastases induced by IL-1 *in vivo* parallels the endothelial cell activation/tumor cell adhesion described *in vitro*.

ENDOTHELIAL CELL ADHESION MOLECULES IN THE METASTATIC PROCESS

Tumor cell extravasation from the blood stream during metastasis is equivalent in many respects to the entry of normal circulating cells into inflammatory tissue.

A major target of inflammatory stimuli is the vascular endothelium which responds with specific morphological and metabolic changes (*Cotran, 1987; Mantovani and Dejana, 1989*). Much of the work on the functional status of endothelial cells derives from studies *in vitro* on the effect of cytokines on cultured endothelial cells. When these cells are activated by IL-1 and TNF, they show increased adhesiveness for

TABLE 1. Principal Molecules involved in tumor cell adhesion to activated endothelial cells

Endothelial cell ligand (family)	Tumor cell receptor (family)	Tumor type	Reference
E-selectin (selectin)	sLewis x; sLewis a and related structures (carbohydrate)	Colon carcinoma Other digestive tract cancers	*Dejana et al., 1992; Lauri et al., 1990; Rice and Bevilacqua, 1989; Walz et al., 1990; Takada et al., 1992*
VCAM-1 (Immunoglobin supergene)	VLA-4 (α4β1) (integrin)	Osteosarcoma Melanoma	*Martin-Padura et al., 1991; Rice and Bevilacqua, 1989*

different subsets of leukocytes (*Osborn, 1990; Zimmerman et al., 1992*). This cell-cell interaction is mediated by the induction or augmented expression of multiple molecules on endothelial cells, such as E-selectin (also known as ELAM-1), intercellular adhesion molecule-1 (ICAM-1) and the vascular cell adhesion molecule-1 (VCAM-1) (*Bevilacqua et al., 1987; Dustin et al., 1986; Osborn et al., 1989*). A logical consequence of these findings was to investigate whether tumor cell adhesion to endothelial cells occurs through recognition mechanisms similar to those used by leukocytes. Indeed, antibodies against these adhesion receptors are able to partially inhibit tumor cell adhesion to cytokine-activated endothelial cells. However, as discussed below, considerable diversity exists with respect to the adhesion pathway preferentially used by different tumor types (Table 1).

E-Selectin

Kinetics studies on the adhesion of human tumor cells on endothelial cells stimulated for various times with IL-1 have shown time-dependent changes in adhesion (*Lauri et al., 1990; Zimmerman et al., 1992*). Specifically, the HT-29 colon carcinoma showed a kinetic of adhesion similar to that described for polymorphonuclear and myeloid cells, that has been shown to be partially inhibited by anti-E-selectin (*Piggot et al., 1991*).

A specific role of E-selectin in the adhesion of HT-29 to activated endothelial cells has been proposed by Rice and Bevilacqua (*Rice and Bevilacqua, 1989*). Extending these studies, Lauri et al., have shown that the adhesion of seven colon carcinoma lines on IL-1-stimulated endothelial cells was E-selectin dependent (*Lauri et al., 1990*). In contrast, anti-E-selectin antibodies did not affect the adhesion of other types of carcinoma, osteosarcomas and melanomas. The counter-receptors for E-selectin have not yet been fully identified, but there is evidence that E-selectin recognizes specific carbohydrate structures such as sialyl-Lewis^x (SLe^x) (*Phillips et al., 1990; Waltz et al., 1990*). We have reported that one monoclonal antibody (MBr8) directed to HT-29 colon carcinoma (*Colnaghi et al.,*), is able to block its adhesion to IL-1-activated endothelial cells (*Dejana et al., 1992*). In contrast, MBr8 did not recognize and did not modify the adhesion of an osteosarcoma cell line (MG63) and polymorphonuclear cells. The relevance of these findings in metastasis is supported by our recent results showing that the augmented retention of HT-29 radiolabelled tumor cells in IL-1-treated mice was inhibited in mice receiving MBr8 antibodies (*Dejana et al., 1992*). MBr8 specifically binds Lewis fucosylated type I and does not recognize SLe^x determinants. These results suggest that E-selectin might be recognized by more than one carbohydrate group on tumor cells or, on the other hand, tumor cells might use more than one carbohydrate group to bind E-selectin. In addition, they support the relevance of developing antibodies that selectively inhibit tumor cell interaction with the vascular wall.

VCAM- 1

Different features of osteosarcoma and melanoma attachment to activated endothelial cells has induced the search for alternative mechanisms that mediate the phenomenon. Rice and Bevilacqua have identified a monoclonal antibody that selectively inhibits the adhesion of melanoma cells, but not colon carcinoma cells, on cytokine-stimulated endothelial cells. This antibody recognizes a receptor, INCAM-1, that by biochemical and functional analysis is similar to VCAM-1 (the endothelial ligand for the very late antigen-4 (VLA-4) integrin) (*Rice and Bevilacqua, 1989*). Studies on a series of melanoma clones have shown differential expression of VLA-4 ($\alpha4\beta1$) integrin that was associated with the ability of the clones to adhere to endothelial cells (*Martin-Padura et al., 1991*). In addition, it has been found that antibodies directed to VLA-4 or VCAM-1 inhibited the adhesion of melanoma clones to IL-1- activated endothelial cells (*Martin-Padura et al., 1991*). While these results strongly support a role for VCAM-1/VLA-4 in melanoma/endothelial cell interactions, it remains to be shown that VLA-4 expression on melanoma cells is associated with a different metastatic behavior, and that anti-VLA-4 can prevent metastasis formation.

ICAM-1

Other endothelial cell adhesion molecules of the immunoglobulin supergene family, such as ICAM-1 and ICAM-2, that serve as endothelial cell surface ligands for leukocyte adhesion seem not to play a role in the adhesion of solid tumors to activated endothelial cells (*Rothlein et al., 1986; Zimmerman et al., 1992*). This is consistent with the lack of the integrin leukocyte function associated antigen-1 (LFA-1) on cells from solid tumors. However, ICAM-1 has been described to play an important role in tumor progression. ICAM-1 expression has been associated with the malignancy of human melanomas (*Johnson et al., 1989; Natali et al., 1990*). On the other hand, ICAM-1 expression on tumor cells may play a role in immune recognition and thus be associated with a favorable prognosis (*Webb et al., 1991*). Recently, elevated levels of soluble ICAM-1 have been identified in the serum of patients with inflammatory and neoplastic disease, including malignant melanomas (*Harning et al., 1991; Seth et al., 1991*). We have found that A375 human melanoma, that expresses ICAM-1 on the cell surface, secretes ICAM-1 after stimulation with the cytokines IL-1, TNF and interferon-γ (*Giavazzi et al., 1992*). The serum from nude mice bearing A375 melanoma tumors was also found to contain soluble human ICAM-1 (*Giavazzi et al., 1992*). The level of ICAM-1 in the serum of mice showed a positive correlation with the tumor burden and was not detected in sera of tumor-free mice or in mice whose primary tumor was surgically removed (unpublished observation). Further investigations are necessary to study the role of ICAM-1 in response to host-derived cytokines and its relevance in tumor progression and metastasis.

CONCLUSIONS

The interaction of tumor cells with the microvasculature is one of the key-steps in tumor dissemination and metastasis. Multiple mechanisms are probably responsible for tumor endothelial cell recognition and several adhesion molecules seem to mediate the adhesion of tumor cells to endothelium. The observation that the treatment of cultured human umbilical endothelial cells with the cytokines, IL-1 or TNF, enhances adhesion of malignant tumor cells suggests that the interaction of tumor cells with endothelium plays an important role in cytokine-induced augmentation of metastases. The adhesion of tumor cells to activated endothelial cells is mediated by specific recognition mechanisms already described for leukocytes, but different adhesion pathways appear to be used by tumor types of different histological origin. A better understanding of the role of these adhesion mechanisms in metastasis is still necessary, but they offer interesting targets for the prognosis and treatment of malignant disease.

ACKNOWLEDGMENTS

Part of this research is supported by the Italian National Research Council (Project ACRO) and by the Italian Association for Cancer Research.

REFERENCES

Auerbach, R., Lu, W.C., Pardon, E., Gumkowski, F., Kaminska, G. and Kaminski, M., 1987, Specificity of adhesion between murine tumor cells and capillary endothelium: An *in vitro* correlate of preferential metastasis *in vivo*. *Cancer Res.* 47:1492.

Bani, M.R., Garofalo, A., Scanziani, E. and Giavazzi, R., 1991, Belardelli, Effect of interleukin-1-beta on metastasis formation in different tumor systems. *J. Nat. Cancer Inst.* 83:119.

Belardelli, F., Ciolli, V., Testa, U., Montesoro, E., Bulgarini, D., Proietti, E., Borghi, P., Sestili, P., Locardi, C., Peschle, C. and Gresser, I., 1989, Anti-tumor effects of interleukin-2 and interleukin-1 in mice transplanted with different syngeneic tumors. *Int. J. Cancer* 44:1108.

Belloni, P.N. and Tressler, R.J., 1989, Microvascular endothelial cell heterogeneity: Interactions with leukocytes and tumor cells. *Cancer Metastasis Rev.* 8:353.

Bevilacqua, M.P., Pober, J.S., Mendrick, D.L., Cotran, R.S. and Gimbrone, M.A., 1987, Identification of an inducible endothelial-leukocyte adhesion molecule. *Proc. Natl. Acad. Sci.* 84:9238.

Colnaghi, M.I., Agresti, R., Ménard, S., Da Dalt, M.G., Cattoretti, G., Andreola, S., Di Fronzo, G., Del Vecchio, M., Verderio, L., Cascinelli, N., Greco, M. and Rilke, F., 1987, Monoclonal antibodies as prognostic indicators of tumor progression. In *Cancer Metastasis*, Prodi, G., Liotta, L.A., Lollini, P.L., Garbisa, S., Gorini, S. and Hellmann, K.(eds), P-319, Plenum Publishing Corporation: New York.

Cotran, R.S., 1987, New roles for the endothelium in inflammation and immunity. *Am. J. Pathol.* 129:407.

Crissman, J.D., Hatfield, J.S., Menter, D.G., Sloane, B. and Honn, K.V., 1988, Morphological study of the interaction of intravascular tumor cells with endothelial cells and subendothelial matrix. *Cancer Res.* 48:4065.

Dejana, E., Bertocchi, F., Bortolami, M.C., Regonesi, A., Tonta, A., Breviario, F. and Giavazzi, R., 1988, Interleukin 1 promotes tumor cell adhesion to cultured human endothelial cells. *J. Clin. Invest.* 82:1466.

Dejana, E., Martin-Padura, I., Lauri, D., Bernasconi, S., Bani, M.R., Garofalo, A., Giavazzi, R., Magnani, J., Mantovani, A. and Menard, S., 1992, ELAM-1-dependent adhesion of colon carcinoma cells to vascular endothelium is inhibited by an antibody to Lewis fucosylated type I carbohydrate chain. *Lab. Invest.* 3:324.

Dustin, M.L., Rothlein, R., Bhan, A.K., Dinarello, C.A. and Springer, T.A., 1986, Induction by IL 1 and interferon-γ: tissue distribution, biochemistry, and function of a natural adherence molecule (ICAM-1). *J. Immunol.* 137:245.

Fidler, J.I., 1990, Critical factors in the biology of human cancer metastasis: twenty-eight G.H.A. Clowes Memorial Award Lecture. *Cancer Res.* 50:6130.

Folkman, J., 1986, How is blood vessel growth regulated in normal and neoplastic tissue? - A. Clowes Memorial Award Lecture. *Cancer Res.* 46:467.

Giavazzi, R., Garofalo, A., Bani, M.R., Abbate, M., Ghezzi, P., Boraschi, D., Mantovani, A. and Dejana, E., 1990, Interleukin-1- induced augmentation of experimental metastases from a human melanoma in nude mice. *Cancer Res.* 50:4771.

Giavazzi, R., Chirivi, R.G.S., Garofalo, A., Rambaldi, A., Hemingway, I., Pigott, R. and Gearing, A.J.H., 1992, Soluble intercellular adhesion molecule-1 is released by human melanoma cells and is associated with tumor growth in nude mice. *Cancer Res.* 52: 2628.

Harning, R., Mainolfi, E., Bystryn, J. C., Henn, M., Merluzzi, V.J. and Rothlein, R., 1991, Serum levels of circulating intercellular adhesion molecule-1 in human malignant melanoma. *Cancer Res.* 51:5003.

Johnson, J.P., Stade, B.G., Holzmann, B., Schwable, W. and Riethmuller, G., 1989, *De novo* expression of intercellular-adhesion molecule-1 in melanoma correlates with increased risk of metastasis. *Proc. Natl. Acad. Sci. USA* 86:641.

Kawaguchi, T. and Nakamura, K., 1986, Analysis of the lodgement and extravasation of tumor cells in experimental models of hematogenous metastasis. *Cancer Metastasis Rev.* 5:77.

Kramer, R.H. and Nicolson, G.L., 1979, Interactions of tumor cells with vascular endothelial cell monolayers: A model for metastatic invasion. *Proc. Natl. Acad. Sci. USA* 76:5704.

Lapis, K., Paku, S. and Liotta, L.A., 1988, Endothelialization of embolized tumor cells during metastasis formation. *Clin. Exp. Metastasis* 6:73.

Lauri, D., Bertomeu, M.C., Orr, F.W., Bastida, E., Sauder, D. and Buchanan, M.R., 1991, Interleukin-1 increases tumor cell adhesion to endothelial cells through an RGD dependent mechanism: *in vitro* and *in vivo* studies. *Clin. Expl. Metastasis* 8:27.

Lauri D., Needham L., Martin-Padura I. and Dejana E., 1990, Tumor cell adhesion to the endothelial cells: ELAM-1 as an inducible adhesive receptor specific for colon carcinoma cells. *J. Natl. Cancer Inst.* 93:1321.

Levine, S. and Saltzman, A., 1990, Lymphatic metastases from the peritoneal cavity are increased in the postinflammatory state. *Invasion Metastasis* 10:281.

Liotta, L.A., 1986, Tumor invasion and metastases-role of the extracellular matrix: Rhoads memorial award lecture. *Cancer Res.* 46:1.

Mantovani, A. and Dejana, E., 1989, Cytokines as communication signals between leukocytes and endothelial cells. *Immunol. Today* 10:370.

Martin-Padura, I., Mortarini, R., Lauri, D., Bernasconi, S., Sanchez-Madrid, F., Parmiani, G., Mantovani, A., Anichini, A. and Dejana, E., 1991, Heterogeneity in human melanoma cell adhesion to cytokine activated endothelial cells correlate with VLA-4 expression. *Cancer Res.* 51:2239.

Murphy, P., Alexander, P., Senior, P.V., Fleming, J., Kirkham, N. and Taylor, I., 1988, Mechanisms of organ selective tumor growth by bloodborn cancer cells. *Br. J. Cancer* 57:19.

Nakamura, S., Nakata, K., Kashimoto, S., Yoshida, H. and Yamada, M., 1986, Antitumor effect of recombinant human interleukin-1 alpha against murine syngeneic tumors. *J. Cancer Res.* 77:767.

Natali, P., Nicotra, M.R., Cavaliere, R., Bigotti, A., Romano, G., Temponi, M. and Ferrone, S., 1990, Differential expression of intercellular adhesion molecule-1 in primary and metastatic melanoma lesions. *Cancer Res.* 50:1271.

Nicolson, G.L., 1988, Organ specificity of tumor metastasis: role of preferential adhesion, invasion and growth of malignant cells at specific secondary sites. *Cancer Metastasis Rev.* 7:143.

Orr, F.W., Adamson, I.Y.R. and Young, L., 1986, Promotion of pulmonary metastasis in mice by bleomycin-induced endothelial injury. *Cancer Res.* 46:891.

Osborn, L., 1990, Leukocyte adhesion to endothelium in inflammation. *Cell* 62:3.

Osborn, L., Hession, C., Tizard, R., Vassallo, C., Luhowskyi, S., Chi-Rosso, G. and Lobb, R., 1989, Direct expression cloning of vascular cell adhesion molecule 1, a cytokine-induced endothelial protein that binds to lymphocytes. *Cell* 59:1203.

Phillips, M.L., Nudelman, E., Gaeta, F.C.A., Perez, M., Singhal, A.K., Hakomori, S.I. and Paulson, J.C., 1990, ELAM-1 mediates cells adhesion by recognition of a carbohydrate ligand syalil-lex. *Science* 250:1130.

Pigot, R., Needham, L.A., Edward, R.M., Walker., C. and Power, C., 1991, Structural and functional studies of the endothelial activation antigen endothelial leukocyte adhesion molecule-1 using a panel of monoclonal antibodies. *J. Immunol.* 147:l3O.

Poste, G. and Fidler, I.J., 1980, The pathogenesis of cancer metastasis. *Nature* 283:139.

Raz, A. and Lotan, R., 1987, Endogenous galactoside-binding lectins: a new class of functional tumor cell surface molecules related to metastasis. *Cancer Met. Rev.* 6:433.

Rice, G.E. and Bevilacqua, M.P., 1989, An inducible endothelial cell surface glycoprotein mediates melanoma adhesion. *Science* 246:1303.

Rice, G.E., Gimbrone, M.A. and Bevilacqua, M.P., 1988, Tumor cell-endothelial interactions. Increased adhesion of human melanoma cells to activated vascular endothelium. *Am. J. Pathol.* 133:204.

Rothlein, R., Dustin, M.L., Marlin, S.D. and Springer, T.A., 1986, A human intercellular adhesion molecule (ICAM-1) distinct from LFA-1. *J. Immunol.* 137:1270.

Ruoslahti, E. and Giancotti, F.G., 1989, Integrins and tumor cell dissemination. *Cancer Cell* 1:119.

Seth, R., Raymond, F.D. and Magkoba, M.W., 1991, Circulating ICAM-1 isoforms: diagnostic prospects for inflammatory and immune disorders. *Lancet* 338:83.

Sugarbaker, E.V., 1979, Patterns of metastasis in human malignancies. *Cancer Biol. Rev.* 2:235.

Takada, A., Ohmori, K., Takahashi, N., Tsuyuoka, K., Yago, A., Kenita, K., Hasegawa, A. and Kannagi, R., 1992, Adhesion of human cancer cells to vascular-endothelium mediated by a carbohydrate antigen, sidyl lewis A. *Biomedical and Biophysical Research Communications*, 179-173.

Van Den Brenk, H.A.S., Burch, W.M., Orton, C. and Sharpington, C., 1973, Stimulation of clonogenic growth of tumor cells and metastases in the lungs by local X-Radiation. *Br. J. Cancer* 27:291.

Walz, G., Aruffo, A., Kolanus, W., Bevilacqua, M.P. and Seed, B., 1990, Recognition by ELAM-1 of the sialyl-lex determinant a myeloid and tumor cells. *Science* 250:1132.

Webb, D.S., Mostowski, H.S. and Gerrard, T.L., 1991, Cytokine-induced enhancement of ICAM-1 expression results in increased vulnerability of tumor cells to monocyte-mediated lysis. *J. Immunol.* 146:3682.

Weiss, L., Orr, F.W. and Honn, K.V., 1989, Interactions between cancer cells and the microvasculature: a rate-regulator for metastasis. *Clin. Expl. Metastasis* 7:127.

Zimmerman, G.A., Prescott, S.M. and McIntyre, T.M., 1992, Leucocyte-endothelial cell interactions. *Immunology Today* 13:93.

III. ALTERED VASCULAR REACTIVITY

ALTERED VASCULAR REACTIVITY IN CORONARY ARTERY DISEASE

Johannes Zanzinger and Eberhard Bassenge

Department of Applied Physiology
University of Freiburg
Freiburg
Germany

INTRODUCTION

The endothelial lining is metabolically very active and synthesizes powerful local hormones (autacoids), which diffuse into the adjacent vasculature and elicit decisive changes in vascular tone and conductance and therefore in organ perfusion. A loss or an impairment of endothelium-dependent control mechanisms becomes particularly important under pathophysiological conditions, such as when endothelial function is disturbed due to atheromatosis, hypertension, diabetes, balloon catheter evoked endothelial denudation, several immunological diseases, including intimal vascular alterations following heart transplantation and other surgical interventions which affect endothelium mediated vascular control. The physiological and pathophysiological aspects of the endothelial autacoid release in the regulation of coronary vascular tone is described in this report with emphasis on the role of the endothelium-derived relaxant factor (EDRF).

SYNTHESIS AND RELEASE OF EDRF

EDRF is very likely identical with nitric oxide (NO) (*Palmer et al., 1987*) or a closely related nitrosyl-compound. It specifically stimulates soluble guanylate cyclase (GC) and thus increases intracellular cGMP-production. In the vasculature, this results in vasodilation after several steps. In platelets, cGMP suppresses intracellular Ca^{2+}-concentrations, activation, adhesion and aggregation (*Busse et al., 1987*), thus maintaining adequate blood fluidity (*Bassenge et al., 1989*). There is a continuous, *basal* release of NO, which is modulated additionally by a second, (receptor-) *stimulated* release, e.g. elicited by acetylcholine or bradykinin, but also by mechanical factors such as blood-flow induced viscous drag or pulsatile stretching imposed on the endothelial lining (for review see *Bassenge and Heusch, 1990*). NO is synthesized in a number of different cell types by the action of a recently discovered enzyme, NO-synthase, present in various isoforms in different cell types as a constitutive (e.g. brain) or inducible (e.g. vasculature) enzyme (*Bredt et al., 1991; Förstermann et al., 1991*). The constitutive NO-synthase in endothelial cells is Ca^{2+}, calmodulin and NADPH-dependent (*Busse and Mülsch, 1990b*). L-arginine acts as a precursor substance and is transformed by various oxidation steps into citrulline, cleaving the guanidino-nitrogen atom to form the NO-radical (*Palmer et al., 1987*). For experimental analyzes, L-arginine can be replaced by a number of biologically inactive arginine analogues like L-mono-methyl-arginine (L-NMMA), L-nitro-arginine (L-NAG), L-amino-arginine (L-NNA) etc.. These stereospecific blockers suppress cellular NO-production, partly by inactivating NO-synthase, partly by interfering with the specific membrane carrier of L-arginine (*Bogle et al., 1991*). However, differential effects of inhibitors of NO-synthesis on basal and stimulated EDRF-release in rabbit hearts recently raised questions whether more than one mechanism of NO-synthesis and/or release might exist (*Lewis, 1992; Smith et al., 1992*).

Flow-induced shear stress acting on the luminal endothelial surface as viscous drag is probably the most important stimulus in the endothelium-dependent control of vascular calibers. This seems to be especially significant in the coronary circulation with its extremely pulsatile nature of flow patterns (flow reversal in systole). Indeed, the pulsatile stretching of the endothelial cell lining has been demonstrated to be another important stimulus for EDRF-release in addition to the viscous drag (*Pohl et al., 1986*). Acute and chronic changes in viscous drag modify substantially EDRF-release from endothelial cells, thereby adjusting vessel calibers in a moment to moment fashion to optimize coronary arterial conductance. For such caliber adjustments, endothelial cells must act as "flow sensors" and must be able to transduce these signals acting mechanically upon the membrane into an intracellular signal associated with altered NO-production.

ALTERATIONS IN CORONARY AND PERIPHERAL HEMODYNAMICS BY ENDOTHELIUM DERIVED NITRIC OXIDE (EDNO/EDRF)

Endothelial NO (EDRF) is continuously synthesized in the coronary bed and released into the coronary effluent (*Kelm and Schrader, 1990*) and is a decisive factor in the control of coronary tone. Without the continuous basal release of EDRF (EDNO) from the endothelial lining, e.g. after blockade of NO-synthesis, an increase in vascular tone (*Bassenge, 1991; Bassenge et al., 1991*) and resistance to flow can be observed in the coronary and in a variety of peripheral vascular beds (*Amezcua et al., 1989; Chu et al., 1990; Chu et al., 1991; Gardiner et al., 1990a; Pohl et al., 1990a; Pohl et al., 1991; Rees et al., 1989; Stewart et al., 1987b; Stewart et al., 1987c*). Effects of EDRF can be suppressed or blocked by binding and inactivating EDRF using hemoglobin (Hb) (*Stewart et al., 1987b; Stewart et al., 1987c*) or by stopping its synthesis by providing biologically inert arginine-analogues such as L-mono-methyl-arginine (L-NMMA) to endothelial cells instead of the biologically active precursor substance for NO generation, namely L-arginine. Blocking EDRF actions by hemoglobin has been shown to cause coronary constriction in isolated perfused hearts and to reduce myocardial perfusion, or when coronary flow is experimentally kept constant, to increase coronary perfusion pressure (*Stewart et al., 1987b; Stewart et al., 1987c*). When the spontaneous endothelial NO-production is suppressed by the administration of the biological inert analogue L-NMMA (*Bassenge, 1991; Pohl et al., 1990b; Pohl et al., 1991*) identical reductions in myocardial perfusion are obtained.

Similar findings have been reported in a number of other peripheral beds (cerebral, mesenteric, renal, skeletal muscle, hindquarters) (*Gardiner et al., 1990a; Pohl et al., 1990a; Pohl et al., 1991*) or in the systemic circulation (*Chu et al., 1990; Chu et al., 1991*).

Following the administration of L-NMMA or a similar analogue like nitro-L-arginine (L-NNA) into the systemic circulation, there is an immediate increase in peripheral resistance and in arterial blood pressure (*Bassenge et al., 1991; Chu et al., 1990; Chu et al., 1991; Rees et al., 1989*). Even when L-NMMA is chronically supplied to rats via the drinking water, there is a long-lasting elevation of peripheral resistance and arterial blood pressure (*Gardiner et al., 1990b*). Thus, the continuous release of EDRF from the endothelial cell lining may have an important modulatory role in suppressing hypertension, vascular alterations and coronary heart disease (*Fuster et al., 1992*). If EDRF release is chronically suppressed, significant hypertension originates, which surprisingly is not easily counterregulated by a number of other biological compensatory mechanisms and reflexes (*Ribeiro et al., 1992*).

To what extent the impaired endothelial function and EDRF release is the cause for the induction of hypertension or whether it reflects the effect of the hypertension-induced endothelial damage, has not yet been analyzed in detail. Experimental hypertension in rats is associated with a substantial reduction in endothelium-mediated vasomotor responses (*Lüscher et al., 1987*) and in hypertensive

46

patients endothelium-dependent regulation of peripheral blood flow (forearm) is likewise reduced (*Linder et al., 1990*).

An increase in peripheral and coronary vascular tone and resistance can also be detected by suppressing the effects of locally released NO acting additionally as an inhibitory neurotransmitter compound in the central nervous system (*Sakuma et al., 1990*). Sympathetic activity (and thus α-adrenergic coronary constriction) has been reported to increase immediately after intracisternal L-NMMA application, augmenting vascular tone and resistance. In favor of this concept, is the fact that intracisternal application of the real biological precursor, L-arginine, reverses this enhancement of sympathetic activity, leading to the cessation of this centrally mediated vasoconstriction. Such a L-NMMA-induced augmentation of sympathetic tone would also tend to limit adequate myocardial oxygen supply.

CORONARY ARTERIOLAR TONE AND MYOCARDIAL PERFUSION ARE ADJUSTED BY ENDOTHELIAL AUTACOIDS

The administration of biological inert arginine analogues, such as L-NNA to beating isolated heart preparations, results in an immediate increase in coronary arteriolar tone (*Amezcua et al., 1989; Pohl et al., 1990b*), which can lower myocardial perfusion under experimental conditions to 50% of the control value (*Pohl et al., 1990b*). Under this experimental condition, there is increased lactate release from the partially ischemic hearts, indicating that a potentially enhanced metabolic stimulation and subsequent release of vasodilator catabolites in this preparation apparently does not compensate adequately for the suppressed endothelium dependent, continuous dilator activity in response to EDRF (*Pohl et al., 1990b*). This finding demonstrates that endothelium dependent control of coronary tone may be as important a modulator mechanism as local metabolic effects and may, under pathophysiological conditions in the presence of an impaired endothelial function, be responsible for an inadequate myocardial perfusion.

An increase in coronary resistance and a simultaneous reduction in maximal coronary conductance during reactive hyperemia was also observed by suppressing the continuous EDRF-mediated coronary dilator effect in chronically instrumented conscious dog preparations (with reflex reactions left intact) after administration of stereospecific NO-blockers (L-NNA) to suppress EDRF-formation (for review see *Bassenge and Heusch, 1990; Bassenge, 1991*). Similar results, using L-NMMA as blocking agent, were obtained by Chu and co-workers (*Chu et al., 1990; Chu et al., 1991*).

ENDOTHELIAL IMPAIRMENT RESULTS IN INADEQUATE CONTROL OF LARGE CORONARY ARTERY CALIBERS

Suppression of EDRF-release in large coronary arteries by L-NMMA or by L-NNA in chronically instrumented conscious dogs leads to a substantial increase in coronary tone marked by a significant diameter reduction even in the presence of an increased arterial perfusion pressure and distending pressure (*Bassenge, 1991; Chu et al., 1990; Chu et al., 1991*).

Following the intracoronary administration of L-NMMA (5mg/kg) into chronically instrumented conscious dogs, there is a substantial suppression of the endothelium-dependent dilator responses to intracoronary acetylcholine (*Bassenge, 1991*). This dilator response is reestablished when L-arginine is administered intracoronarily (30 mg/kg). Similarly suppressed is the flow-dependent dilator response (FDD) under L-NMMA, a response which is also dependent on an unimpaired endothelial function through shear stress evoked EDRF-release. In contrast, the endothelium-independent dilator responses to nitroglycerin are not reduced. After i.c. administration of L-arginine, the FDD response was reestablished. How much of this L-NMMA-induced suppression of FDD response may result from the slightly reduced maximal reactive hyperemic response after intracoronary L-NMMA cannot be exactly quantified under this *in vivo* conditions.

These findings demonstrate that also large coronary conductance arteries are under the control of a continuous basal EDRF-release which tends to augment coronary conductance and to counterbalance the continuous myogenic constrictor impulses (*Pohl et al., 1990b*), which - mainly in the arteriolar resistance section - reduce coronary conductivity and myocardial perfusion.

IMPAIRED EDRF-FORMATION/RELEASE AND ENHANCED INACTIVATION OF EDRF IN CORONARY ARTERY DISEASE

In various ischemic forms of coronary artery disease (CAD) with impaired endothelial function, i.e. hypercholesterolemia (*Creager et al., 1990*), atherosclerosis (*Harrison et al., 1987a; Harrison et al., 1987b*), hypertension-induced vascular damage (*Panza et al., 1990*), balloon catheter induced endothelial denudation (*Fischell et al., 1989*), diabetes (*Saenz de Tejada et al., 1989*), reperfusion damage (*Lefer and Lefer, 1991; Stewart et al., 1988*), coronary spasm (*Chesebro et al., 1989; Fischell et al., 1989; Nagasawa et al., 1989*), subarachnoid hemorrhage induced vasospasm (*Hongo et al., 1988; Nakagomi et al., 1987*) the autacoid- (and EDRF-) release is depressed or the released NO-radical immediately inactivated by hemoglobin or by oxygen derived radicals before affecting the vasculature (*Mügge et al., 1992*). The impaired and deficient EDRF release favors an enhanced vasoconstrictor tone especially under the condition of an insufficient mechanically or receptor-stimulated NO-release in large feed arteries (e.g. coronaries). This becomes particularly obvious when endothelial NO-production is reduced or absent upon stimulation by various factors which initiate endothelial release of autacoids. Therefore, the combined net vasomotor effect of circulating agonists and endothelial autacoids may shift gradually to reduced vasodilation, or even to an excessive vasoconstrictor tone. Thus, exogenous acetylcholine intracoronarily injected causes dose-dependent dilations in patients without atheromatosis and coronary risk factors, in the presence of CAD however, it causes dose dependent constrictor responses (*Ludmer et al., 1986; Zeiher et al., 1989*). This may result in an inadequate blood supply and thus in ischemic damage. A diminished dilator response of the resistance arteries, especially in the presence of multiple risk factors, can add to this dysregulation. Such a tendency of inadequate dilator capacity has been observed even in the course of normal aging in humans (*Zeiher et al., 1992*) in the absence of CAD.

The stimulation of immunization processes in patients with heart transplantation leads likewise to significant functional endothelial impairment; thus, it is not surprising that, in these particular patients exclusively, coronary constrictions are observed in the transplanted vessels upon intracoronary acetylcholine test injections in contrast to normal controls (*Nellessen et al., 1988*). Probably such reactions can be used to test endothelial integrity and function, which also seems to deteriorate with age independent of additional risk factors (*Zeiher et al., 1992*).

The equilibrium between vasoconstrictor and vasodilator mechanisms for the maintenance of an adequate coronary tone is disturbed in CAD. This results in inappropriate coronary vasoconstriction, thereby further aggravating CAD, which could be demonstrated in *in-vitro*-experiments, in animal models and recently in a number of clinical studies on patients.

a) In hypercholesterolemia the augmented low density lipoprotein fraction can act as a scavenger for EDRF released from endothelial cells resulting in an accelerated inactivation of EDRF similar to the well described action of hemoglobin as inactivator of EDRF/NO (formation of NO-Hb) or oxygen derived free radicals (*Mügge et al., 1992; Stewart et al., 1988*). Under clinical conditions this scavenger effect is probably of lesser importance.

b) There is a substantial reduction of EDRF/NO release from endothelial cells during hypercholesterolemia, atheromatosis and arteriosclerosis. Part of this is explained by endothelial impairment and cell necrosis. However, the mechanisms and metabolic alterations involved when endothelial morphology is still well preserved, could not clearly be identified so far. *Mügge et al., 1992* point out that the enhanced generation of oxygen derived free radicals in atherosclerotic arteries contributes significantly to the endothelial dysfunction and depressed EDRF-activity.

c) Direct effects of LDL on smooth muscle tone of various vascular preparations are probably based mainly on oxidatively modified LDL, since LDL under *in vitro* conditions undergoes spontaneous oxidation to form lipid peroxides and ox-LDL. In a number of papers, this was recently demonstrated (*Galle et al., 1990; Simon et al., 1990*). Increased plasma LDL-levels result in an augmented activation of macrophages, favoring an increased formation of ox-LDL and its storage and deposition in the vascular wall. In addition, endothelial cells themselves can oxidize LDL. In atheromatosis and atherosclerosis, ox-LDL can be demonstrated in the vascular wall. In *in vitro* experiments with perfused vessel segments, application of ox-LDL results in a substantial potentiation of agonist-induced constrictions (e.g. by catecholamines, serotonin, angiotensin etc.), which is not altered by the absence or presence of an endothelial lining, but substantially reduced in the presence of Ca^{2+}-antagonists. Similar results were obtained when arteries were perfused with cholesterol-phospholipid-containing media and then exposed to adrenergic stimulation. In patients with CAD, this becomes also evident when mental stress precipitates myocardial ischemia (*Yeung et al., 1991*). There was a marked potentiation with an absolute dependence on extracellular Ca^{2+} (*Broderick et al., 1989*). Comparable results were obtained in the coronary system of hypercholesterolemic conscious dogs (*Rosendorf et al., 1981*). An increased transmembrane Ca^{2+}-influx following intracoronary H_1-stimulation in miniature swine with balloon catheter induced regional intimal thickening, however, in the presence of a regrown (obviously dysfunctional) endothelial lining, was apparently the reason for the induction of focal coronary spasm three months after denudation (*Yamamoto et al., 1987*).

d) A diminished sensitivity of vessel segments from cholesterol fed and atheromatotic animals to Ach (receptor-dependent) stimulation EDRF/NO-superfusion could be demonstrated. Recently, hypercholesterolemia, but not hypertension without left ventricular hypertrophy, was identified to cause impaired endothelium-dependent dilation in CAD-patients (*Zeiher et al., 1992*). The mechanism responsible for this desensitization could not be identified yet. Subintimal thickening as an additional diffusion barrier to EDRF/NO could be excluded (*Harrison et al., 1987a*).

e) Increased oxLDL-levels, but not native LDL, stimulate the expression and release of endothelin, which may enhance constrictive coronary tone and sensitize the coronaries for other constrictive stimuli like angiotensin or serotonin (*Boulanger et al., 1992*).

ROLE OF ENDOTHELIN IN CAD

Though there are a large number of recent studies on the cardiovascular effects of endothelin (for review see *Luscher, 1991*), little information exists on the pathophysiological significance of endothelin released from endothelial cells during CAD. Intracoronary administered endothelin causes vasoconstriction mainly via the smooth muscle ET_A receptor (*Hom et al., 1992*). These constrictions reach excessive magnitudes and can induce severe myocardial ischemia resulting in acute heart failure. This potential for severe constriction has been associated with the induction of coronary spasm. The endothelial ET_B receptor is known to mediate prostacyclin and EDRF-release form the endothelium (*Sakurai et al., 1990*), an effect which lacks when endothelial damage is present in CAD. These conclusions are similar to the ones drawn in a large number of reports in the last 20 years on the cause of coronary spasm in conjunction with α-sympathetic constriction, unbalanced release of PGI_2/TxA_2, impaired EDRF-release, endothelial dysfunction and recently excessive release of endothelin. However, any conclusive evidence for a role of endothelin as the single or main cause for spasm is still missing. In patients with atherosclerosis and hyperlipidemia, circulating levels of endothelin are increased (*Arendt et al., 1990*), a finding which is supported by studies on endothelial cells showing increased expression of preproendothelin mRNA in response to oxLDL's (*Boulanger et al., 1992*). Most recently, vasoconstrictions to Sarafotoxin S6C, a selective ET_B-receptor agonist, has been shown *in vivo* (anaesthetized rats) (*Clozel et al., 1992*) indicating indirect constrictive responses also to ET_B-stimulation. Thus, in CAD, the constrictive effects of endothelin may be significantly potentiated.

SUBSTITUTION OF EDRF BY EXOGENOUS NO CLEAVED FROM NITROVASODILATORS

In CAD, nitrovasodilators are used to substitute for a diminished endothelial EDRF/NO-production since nitrovasodilators have been shown, after being processed in a number of poorly analyzed metabolic steps, to finally yield NO (*Feelisch and Noack, 1987*). This, after stimulation of cGMP-production, can initiate relaxation and compensate for insufficient endothelial EDRF/NO production e.g. in atherosclerotic vessel sections.

It recently became obvious that such a substitution is probably effective, especially in vascular segments, which show - due to endothelial impairment or denudation - reduced EDRF-production and thus inadequate cGMP-production and relaxation (*Bassenge and Stewart, 1988*). There is, both *in vitro* (*Bassenge and Stewart, 1988*) and *in vivo*, evidence (*Rafflenbeul et al., 1989*) that nitrovasodilator-induced dilation is more pronounced in endothelium impaired or denuded sections as compared to endothelium-intact segments. *In vitro* denuded sections relax about three times more effectively to nitrovasodilators (*Bassenge, 1989; Busse et al., 1989*), and similar responses have been observed in atherosclerotic, but still compliant, vessel sections in CAD patients in *in vivo* studies (*Rafflenbeul et al., 1989*).

This surprising increase in sensitivity may not only be explained by the augmentation of (the absolute) cGMP-levels in the vasculature but also by the rate of increase of cGMP-levels (relative changes) upon a sudden increase in intravascular nitrovasodilator concentrations (*Bassenge, 1991*).

The continuous application of exogenous nitrovasodilators leads to "tolerance", both in the coronary arteries and in the venous system (*Stewart et al., 1986; Stewart et al., 1987a*), probably by a limitation of the biotransformation from nitrovasodilators into the active compound NO. This results in diminished venodilation and less reduction of preload during nitrate administration (*Münzel et al., 1990*). Thus, there is a smaller decrease of ventricular wall tension and myocardial oxygen consumption. A withdrawal of the nitrate therapy in a tolerant state results in rebound coronary constrictions in dogs which are not related to an activated renin-angiotensin system (*Münzel et al., 1992*).

In such a state of tolerance, the vascular responses to "endogenous nitrate" EDRF produced and released by the endothelial cell lining are not reduced (*Stewart et al., 1987a*). The dilator responses to an exogenous nitrate, namely nitroglycerin, are substantially suppressed during tolerance, the responses to the "endogenous nitrate" EDRF are not affected after the induction of tolerance. Similarly the response to an endogenous stimulator of NO-release, the mechanical shear stress exerted by the blood flow as viscous drag upon the endothelial lining is not suppressed during tolerance. Thus, a reduced sensitivity of the vasculature cannot be demonstrated, but the bioconversion of the exogenous nitrates to NO seems to be limited during tolerance e.g. by a change in cytochrome-P450 like activity (*McDonald and Bennett, 1990; Servent et al., 1989*).

SIDE EFFECTS OF "MANIPULATING" EDRF/NO-PRODUCTION IN CAD

Recently, it was shown that in cytokine and endotoxin-induced shock the specific constitutive enzyme NO-synthase in endothelial cells is excessively expressed as inducible enzyme in the vasculature (*Busse and Mülsch, 1990a*) accounting for an uncontrolled NO-release and excessive vasodilation during various shock conditions. The therapeutical application of various stereospecific inhibitors of the arginine dependent NO-production may turn out to be a useful therapeutic tool in such states of uncontrolled and excessive vasodilator activity. The induction of NO-synthase in the vasculature can be suppressed by antiinflammatory cortico-steroids (*Radomski et al., 1990*). Whether the beneficial effects of feeding fish oil to counterbalance atherosclerosis induced vasomotor abnormalities can be attributed to improved endothelial autacoid production (*Shimokawa et al., 1987; Shimokawa and Vanhoutte, 1989a; Shimokawa and Vanhoutte, 1989b*) remains to be demonstrated.

FUTURE TRENDS FOR THE IMPROVEMENT
OF ANTIISCHEMIC TREATMENT

Antiischemic treatment in patients with CAD in the next years will probably include methods for the preservation of endothelial autacoid release, e.g. by the suppression of noxious (e.g. atherogenic) factors. In addition, measures may be taken to increase intracellular arginine concentrations by supplementing exogenous L-arginine and stimulating NO-synthase in the endothelial lining to augment EDRF-/NO-formation in affected vessel sections. Under pathophysiological conditions, NO-synthase is excessively induced and expressed in vascular smooth muscle by endotoxins and cytokines in a variety of shock conditions, which can be suppressed and normalized by corticosteroids (*Radomski et al., 1990*; for review see *Moncada et al., 1991*).

Finally, the substitution therapy of endothelial autacoid production (such as nitrovasodilator administration) will probably be continued and improved. An important task for the future will be to overcome the development of tolerance during continued nitrovasodilator treatment. This may be achieved by administering agents that directly release NO (which do not need several intermediate metabolic steps in order to yield NO, (*Bassenge et al., 1992; Bohn et al., 1991*) or by the interposition of other dilator principals, such as Ca^{2+}-antagonists or K^+-channel openers. With increasing extent, nitrate therapy leads to activation of counterregulatory factors, including stimulation of the adrenergic and of the renin angiotensin system (*Bassenge and Zanzinger, 1992; Parker et al., 1991*). Thus, there may be a beneficial role for angiotensin converting enzyme inhibitors in the prevention or amelioration of nitrate tolerance (*Katz et al., 1991*).

REFERENCES

Amezcua, J.L., Palmer, R.M.J., DeSouza, B.M. and Moncada, S., 1989, Nitric oxide synthesized from L arginine regulates vascular tone in the coronary circulation of the rabbit. *Br. J. Pharmacol.* 97:1019-1024.

Arendt, R.M., Wilbert-Lamper, U. and Heucke, L., 1990, Increased plasma endothelin in patients with hyperlipoproteinemia and stable or unstable angina (abstract). *Circulation* 82:4.

Bassenge, E. and Stewart, D.J., 1988, Interdependence of pharmacologically-induced and endothelium-mediated coronary vasodilation in antianginal therapy. *Cardiovasc. Drugs* 2:27-34.

Bassenge, E., 1989, Flow-dependent regulation of coronary vasomotor tone. *Eur. Heart J.* 10 (Suppl F): 22-27.

Bassenge, E., Busse, R. and Pohl, U., 1989, Hemmung der Thrombozytenaggregation und -adhäsion durch EDRF und deren pathophysiologische Bedeutung (Inhibition of platelet-aggregation and-adhesion by EDRF: pathophysiological significance). *Z. Kardiol.* 78(Suppl 6):54-58.

Bassenge, E. and Heusch, G., 1990, Endothelial and neuro-humoral control of coronary blood flow in health and disease. *Rev. Physiol. Biochem. Pharmacol.* 116:77-165.

Bassenge, E., 1991, Endothelium-mediated regulation of coronary tone. *Basic Res. Cardiol.* 86 (Suppl.2):69-76.

Bassenge, E., Münzel, T. and Huckstorf, C., 1991, Endothelium-derived nitric oxide predominates in the regulation of coronary tone and myocardial perfusion. *FASEB J.* 5(Part II):A1393.

Bassenge, E., Huckstorf, C. and Münzel, T., 1992, Pronounced coronary dilation by a new NO-releasing sydnonimine derivate during 5 day infusion. *FASEB J.* 6 (Part I):A1579.

Bassenge, E. and Zanzinger, J., 1992, Nitrates in different vascular beds, nitrate tolerance, and interactions with endothelial function. *Am. J. Cardiol.* 69:23B-29B.

Bogle, R.G., Moncada, S., Pearson, J.D. and Mann, G.E., 1991, Identification of sensitive inhibitors of arginine transport and nitric oxide synthase in vascular endothelial cells. In: Moncada, S., Marletta, M.A., Hibbs, J. (eds); Proc. 2 Int. Meeting: Biology of Nitric Oxide. London.

Bohn, H., Beyerle, R., Martorana, P.A. and Schönafinger, K., 1991, CAS 936, a novel sydnonimine with direct vasodilating and nitric oxide-donating properties: Effects on isolated blood vessels. *J. Cardiovasc. Pharmacol.* 18:522-527.

Boulanger, C.M., Tanner, F.C., Béa, M.L., Hahn, A.W.A., Werner, A. and Lüscher, T.F., 1992, Oxidized low density lipoproteins induce mRNA expression and release of endothelin from human and porcine endothelium. *Circ. Res.* 70:1191-1197.

Bredt, D.S., Hwang, P.M., Glatt, C.E., Lowenstein, C., Reed, R.R. and Snyder, S.H., 1991, Cloned and expressed nitric oxide synthase structurally resembles cytochrome P-450 reductase. *Nature* 351:714-718.

Broderick, R., Bialecki, R. and Tulenko, T., 1989, Cholesterol-induced changes in rabbit arterial smooth muscle sensitivity to adrenergic stimulation. *Am. J. Physiol.* 257:H170-H178.

Busse, R., Lückhoff, A. and Bassenge, E., 1987, Endothelium-derived relaxant factor inhibits platelet activation. *Naunyn-Schmiedebergs Arch. Pharmacol.* 336:566-571.

Busse, R., Pohl, U., Mülsch, A. and Bassenge, E., 1989, Modulation of the vasodilator action of SIN-1 by the endothelium. *J. Cardiovasc. Pharmacol.* 14 (Suppl.11):S81-S85.

Busse, R. and Mülsch, A., 1990a, Calcium-dependent nitric oxide synthesis in endothelial cytosol is mediated by calmodulin. *FEBS Letters* 265:133-136.

Busse, R. and Mülsch, A., 1990b, Induction of nitric oxide synthase by cytokines in vascular smooth muscle cells. *FEBS Letters* 275:87-90.

Chesebro, J.H., Fuster, V. and Webster, M.W.I., 1989, Endothelial injury and coronary vasomotion. *J. Am. Coll. Cardiol.,* 14:1191-1192.

Chu, A., Chambers, D.E., Lin, C.C., Kuehl, W.D. and Cobb, F.R., 1990, Nitric oxide modulates epicardial coronary basal vasomotor tone in awake dogs. *Am. J. Physiol.* 258:H1250-H1254.

Chu, A., Chambers, D.E., Lin, C.C., Kuehl, W.D. and Palmer, R.M.J., 1991, Effects of inhibition of nitric oxide formation on basal vasomotion and endothelium-dependent responses of the coronary arteries in awake dogs. *J. Clin. Invest.* 87:1964-1968.

Clozel, M., Gray, G.A., Breu, V., Löffler, B.M. and Osterwalder, R., 1992, The endothelin ETB receptor mediates both vasodilation and vasoconstriction *in vivo. Biochem. Biophys. Res. Commmun.* 186:867-873.

Creager, M.A., Cooke, J.P., Mendelsohn, M.E., Gallagher, S.J., Coleman, S.M., Loscalzo, J. and Dzau, V.H., 1990, Impaired vasodilation of forearm resistance vessels in hypercholesterolemic humans. *J. Clin. Invest.* 86:228-234.

Feelisch, M. and Noack, E., 1987, Correlation between nitric oxide formation during degradation of organic nitrates and activation of guanylate cyclase. *Eur. J. Pharmacol.* 139:19-30.

Fischell, T.A., Nellessen, U., Johnson, D.E. and Ginsburg, R., 1989, Endothelium-dependent arterial vasoconstriction after balloon angioplasty. *Circulation* 79:899-910.

Förstermann, U., Schmidt, H.H.H.W., Pollock, J.S., Sheng, H., Mitchell, J.A., Warner, T.D., Nakane, M. and Murad, F., 1991, Isoforms of nitric oxide synthase: characterization and purification from different cell types. *Biochem. Pharmacol.* 42:1849-1857.

Fuster, V., Badimon, L., Badimon, J.J. and Chesebro, J.H., 1992, Mechanisms of disease - the pathogenesis of coronary artery disease and the acute coronary syndromes. *N. Engl. J. Med.* 326:242-250.

Galle, J., Luley, C. and Bassenge, E., 1990, Hypercholesterolemia causes hyperresponsiveness to serotonin in rabbit aortae, associated with lipid peroxide formation but not with attenuation of EDRF-mediated vasodilation. *Eur. Heart J.* 11(Suppl):265.

Gardiner, S.M., Compton, A.M., Bennett, T., Palmer, R.M.J. and Moncada, S., 1990a, Control of regional blood flow by endothelium-derived nitric oxide. *Hypertension* 15:486-492.

Gardiner, S.M., Compton, A.M., Bennett, T., Palmer, R.M.J. and Moncada, S., 1990b, Regional hemodynamic changes during oral ingestion of ng-monomethyl-Larginine or ng-nitro-Larginine methyl ester in conscious brattleboro rats. *Br. J. Pharmacol.* 101:10-12.

Harrison, D.G., Armstrong, M.L., Freiman, P.C. and Heistad, D.D., 1987a, Restoration of endothelium-dependent relaxation by dietary treatment of atherosclerosis. *J. Clin. Invest.* 80:1808-1811.

Harrison, D.G., Freiman, P.C., Armstrong, M.L., Marcus, M.L. and Heistad, D.D., 1987b, Alterations of vascular reactivity in atherosclerosis. *Circ. Res.* 61(Suppl.II):II-74-II-80.

Hom, G.J., Touhey, B. and Rubanyi, G.M., 1992, Effects of intracoronary administration of endothelin in anesthetized dogs - comparison with bay k-8644 and U-46619. *J. Cardiovasc. Pharmacol.* 19:194-200.

Hongo, K., Kassell, N.F., Nakagomi, T., Sasaki, T., Tsukahara, T., Ogawa, H., Vollmer, D.G. and Leh mann, R.M., 1988, Subarachnoid hemorrhage inhibition of endothelium-derived relaxing factor in rabbit basilar artery. *J. Neurosurg.* 69:247-253.

Katz, R.J., Levy, W.S., Buff, L. and Wasserman, A.G., 1991, Prevention of nitrate tolerance with angiotensin converting enzyme inhibitors. *Circulation* 83:1271-1277.

Kelm, M. and Schrader, J., 1990, Control of coronary vascular tone by nitric oxide. *Circ. Res.* 66:1561-1575.

Lefer, A.M. and Lefer, D.J., 1991, Endothelial dysfunction in myocardial ischemia and reperfusion: role of oxygen-derived free radicals. *Basic Res. Cardiol.* 86(Suppl.2):109-116.

Lewis, M.J., 1992, Nitric oxide and regulation of coronary vascular tone. *Cardiovasc. Res.* 26:555-556.

Linder, L., Kiowski, W., Buhler, F.R. and Luscher, T.F., 1990, Indirect evidence for release of endothelium-derived relaxing factor in human forearm circulation *in vivo* - blunted response in essential hypertension. *Circulation* 81:1762-1767.

Ludmer, P.L., Selwyn, A.P., Shook, T.L., Wayne, R.R., Mudge, G.H., Alexander, R.W. and Ganz, P., 1986, Paradoxical vasoconstriction induced by acetylcholine in atherosclerotic coronary arteries. *N. Engl. J. Med.* 315:1046-1051.

Lüscher, T., Raij, L. and Vanhoutte, P.M., 1987, Endothelium-dependent vascular responses in normotensive and hypertensive Dahl rats. Hypertension, 9:157-163.

Luscher, T.F., 1991, Endothelin. *J. Cardiovasc. Pharmacol.* 18:S15-S22.

McDonald, B.J. and Bennett, B.M., 1990, Cytochrome P-450 mediated biotransformation of organic nitrates. *Can. J. Physiol. Pharmacol.* 68:1552-1557.

Moncada, S., Palmer, R.M.J. and Higgs, E.A., 1991, Nitric oxide - physiology, pathophysiology, and pharmacology. *Pharmacol. Rev.* 43:109-142.

Mügge, A., Elwell, J.H., Peterson, T.E., Hofmeyer, T.G., Heistad, D.D. and Harrison, D.G., 1992, Chronic treatment with polyethylene-glycolated superoxide dismutase partially restores endothelium-dependent vascular relaxations in cholesterol-fed rabbits. *Circ. Res.* (in press).

Münzel, T., Just, H. and Bassenge, E., 1990, Inhibition of EDRF release by N-nitro-L-arginine induces potent venoconstriction *in vivo*. *Circulation* 82(Suppl.III):III-456.

Münzel, T., Zanzinger, J., Bassenge, E., 1992, Sudden withdrawal of chronic glyceryl trinitrate infusion causes rebound constriction of large coronary arteries. *Eur Heart J* 13(Suppl.):166(abstract).

Nagasawa, K., Tomoike, H., Hayashi, Y., Yamada, A., Yamamoto, T. and Nakamura, M., 1989, Intramural hemorrhage and endothelial changes in atherosclerotic coronary artery after repetitive episodes of spasm in X-ray irradiated hypercholesterolemic pigs. *Circ. Res.* 65:272-282.

Nakagomi, T., Kassell, N.F., Sasaki, T., Fujiwara, S., Lehmann, R.M., Joshita, H., Nazar, G.B. and Torner, J.C., 1987, Effect of subarachnoidal hemorrhage on endothelium-dependent vasodilation. *J. Neurosurg.* 66:915-923.

Nellessen, U., Lee, T.C., Fischell, T.A., Ginsburg, R., Masuyama, T., Alderman, E.L. and Schroeder, J.S., 1988, Effects of acetylcholine on epicardial coronary arteries after cardiac transplantation without angiographic evidence of fixed graft narrowing. *Am. J. Cardiol.* 62:1093-1097.

Palmer, R.M.J., Ferrige, A.G. and Moncada, S., 1987, Nitric oxide release accounts for the biological activity of endothelium- derived relaxing factor. *Nature* 327:524-526.

Panza, J.A., Quyyumi, A.A., Brush, J.E. and Epstein, S.E., 1990, Abnormal endothelium-dependent vascular relaxation in patients with essential hypertension. *N. Engl. J. Med.* 323:22-27.

Parker, J.D., Farrell, B., Fenton, T., Cohanim, M. and Parker, J.O., 1991, Counter-regulatory responses to continuous and intermittent therapy with nitroglycerin. *Circulation* 84:2336-2345.

Pohl, U., Busse, R., Kuon, E. and Bassenge, E., 1986, Pulsatile perfusion stimulates the release of endothelial autacoids. *J. Appl. Cardiol.* 1:215-235.

Pohl, U., Herlan, K., Huang, A. and Bassenge, E., 1990a, Inhibitory effects of EDRF-mediated flow-induced dilation on myogenic vasoconstriction (abstract). *Pflügers Arch.* 415(Suppl.1):R62.

Pohl, U., Lamontagne, D., Bassenge, E. and Busse, R., 1990b, EDRF augments coronary conductivity through attenuation of myogenic autoregulation. *Pflügers Arch.* 415(Suppl.1):R62.

Pohl, U., Herlan, K., Huang, A. and Bassenge, E., 1991, EDRF-mediated shear-induced dilation opposes my-ogenic vasoconstriction in small rabbit arteries. *Am. J. Physiol.* 261:H2016-H2023.

Radomski, M.W., Palmer, R.M.J. and Moncada, S., 1990, Glucocorticoids inhibit the expression of an inducible, but not the constitutive, nitric oxide synthase in vascular endothelial cells. *Proc. Natl. Acad. Sci. U. S. A.* 87:10043-10047.

Rafflenbeul, W., Bassenge, E. and Lichtlen, P., 1989, Competition between endothelium-dependent and nitroglycerin-induced coronary vasodilation (Konkurrenz zwischen endothelabhängiger und Nitroglycerin-induzierter koronarer Vasodilatation). *Z. Kardiol.* 78(Suppl 2):45-47.

Rees, D.D., Palmer, R.M. and Moncada, S., 1989, Role of endothelium-derived nitric oxide in the regulation of blood pressure. *Proc. Natl. Acad. Sci. USA.* 86:3375-3378.

Ribeiro, M.O., Antunes, E., de Nucci, G., Lovisolo, S.M., Zatz, R., 1992, Chronic inhibition of nitric oxide synthesis. A new model of arterial hypertension. *Hypertension* 20:298-303.

Rosendorf, C., Hoffman, J.I.E., Verrier, E.D., Rouleau, J. and Boerboom, L.E., 1981, Cholesterol potentiates the coronary artery response to norepinephrine in anesthetized and conscious dogs. *Circ. Res.* 48:320-329.

Saenz de Tejada, I., Goldstein, I., Azadzoi, K., Krane, R.J. and Cohen, R.A., 1989, Impaired neurogenic and endothelium-mediated relaxation of penile smooth muscle from diabetic men with impotence. *N. Engl. J. Med.* 320:1025-1030.

Sakuma, I., Togashi, H., Yoshioka, M., Kobayashi, T., Saito, H., Yasuda, H., Gross, S.S. and Levi, R., 1990, Effects of intravenous and intracisternal administration of NG-monomethyl-L-arginine on renal sympathetic nerve activity in anaesthetized rats. In: Moncada, S., Higgs, E.A. (eds); Nitric oxide from L-arginine: a bioregulatory system. *Elsevier, Amsterdam* 481-482.

Sakurai, T., Yanagisawa, M., Takuwa, Y., Miyazaki, H., Kimura, S., Goto, K. and Masaki, T., Cloning of a cDNA encoding a non-isopeptide-selective subtype of the endothelin receptor. *Nature* 348:732-735.

Servent, D., Delaforge, M., Ducrocq, C., Mansuy, D. and Lenfant, M., 1989, Nitric oxide formation during microsomal hepatic denitration of glyceryl trinitrate: involvement of cytochrome P-450. *Biochem. Biophys. Res. Commun.* 163:1210-1216.

Shimokawa, H., Aarhus, L.L. and Vanhoutte, P.M., 1987, Porcine coronary arteries with regenerated endothelium have a reduced endothelium-dependent responsiveness to aggregating platelets and serotonin. *Circ. Res.,* 61:256-270.

Shimokawa, H. and Vanhoutte, P.M., 1989a, Dietary omega-3 fatty acids and endothelium-dependent relaxations in porcine coronary arteries. *Am. J. Physiol.* 256:H968-H973.

Shimokawa, H. and Vanhoutte, P.M., 1989b, Impaired endothelium-dependent relaxation to aggrega-t ing plate-lets and related vasoactive substances in porcine coronary arteries in hypercholester-olemia and atherosclerosis. *Circ. Res.* 64:900-914.

Simon, B.C., Cunningham, L.D. and Cohen, R.A., 1990, Oxidized low density lipoproteins cause contraction and inhibit endothelium-dependent relaxation in the pig coronary artery. *J. Clin. Invest.* 86:75-79.

Smith, R.E.A., Palmer, R.M.J., Bucknall, C.A. and Moncada, S., 1992, Role of nitric oxide synthesis in the regulation of coronary vascular tone in the isolated perfused rabbit heart. *Cardiovasc. Res.* 26:508-512.

Stewart, D.J., Elsner, D., Sommer, O., Holtz, J. and Bassenge, E., 1986, Altered spectrum of nitrogly-cerin action in long term treatment: nitroglycerin-specific venous tolerance with maintenance of arterial vasodepressor potency. *Circulation* 74:573-582.

Stewart, D.J., Holtz, J. and Bassenge, E., 1987a, Long-term nitroglycerin treatment: effect on direct and endothelium-mediated large coronary artery dilation in conscious dogs. *Circulation* 75:847-8562.

Stewart, D.J., Holtz, J., Pohl, U. and Bassenge, E., 1987b, Balance between endothelium-mediated dilator and direct constrictor actions of serotonin on resistance vessels in the isolated rabbit heart. *Eur. J. Pharmacol.* 143:131-134.

Stewart, D.J., Münzel, T. and Bassenge, E., 1987c, Reversal of acetylcholine-induced coronary resis-tance vessel dilation by hemoglobin. *Eur. J. Pharmacol.* 136:239-242.

Stewart, D.J., Pohl, U. and Bassenge, E., 1988, Free radicals inhibit endothelium-dependent dilation in the coronary resistance bed. *Am. J. Physiol.* 255:H765-H769.

Yamamoto, Y., Tomoike, H., Egashira, K., Kobayashi, T., Kawasaki, T. and Nakamura, M., 1987, Pathogenesis of coronary artery spasm in miniature swine with regional intimal thickening after balloon denudation. *Circ. Res.* 60:113-121.

Yeung, A.C., Vekshtein, V.I., Krantz, D.S., Vita, J.A., Ryan, T.J., Ganz, P. and Selwyn, A.P., 1991, The effect of atherosclerosis on the vasomotor response of coronary arteries to mental stress. *N. Engl. J. Med.* 325:1551-1556.

Zeiher, A.M., Drexler, H., Wollschlaeger, H., Saurbier, B. and Just, H., 1989, Coronary vasomotion in response to sympathetic stimulation in humans - importance of the functional integrity of the endothelium. *J. Am. Coll. Cardiol.* 14:1181-1190.

Zeiher, A.M., Drexler, H., Saurbier, B., Schächinger, V. and Just, H., 1992, Effects of risk factors for CAD on endothelium-dependent modulation of coronary blood flow in humans (abstract). *Eur. Heart J.* 13(Suppl.):7.

Zeiher, A.M., Drexler, H., Saurbier, B., Schächtinger, V., Bitter, M. and Just, H., 1992, Modulation of coronary blood flow by the endothelium in humans: effects of age, atherosclerosis, hyper-cholesterolemia, and hypertension. *J. Clin. Invest.* (submitted).

CORONARY ARTERY AND CORONARY ARTERY BYPASS GRAFT DISEASE:

ROLE OF THE ENDOTHELIUM AND VASCULAR SMOOTH MUSCLE

Thomas F. Lüscher

Cardiology
Division of Cardiovascular Research
University Hospital
Bern
Switzerland

INTRODUCTION

Coronary artery disease and its sequelae, such as myocardial infarction and sudden death, is one of the most important causes of morbidity and mortality in Western countries. The pathophysiology of *coronary artery disease* is not fully understood, but three main alterations have been delineated: (1) Increased vasoconstrictor responses in the coronary circulation; (2) augmented platelet-vessel wall interaction and (3) proliferative responses in the blood vessel wall leading to narrowing of major epicardial coronary arteries. All three factors work in concert to cause myocardial ischemia and eventually infarction and death.

Research on the vascular biology of coronary artery disease suggests that alterations in endothelial function and that of vascular smooth muscle are important factors in those processes (*Lüscher and Vanhoutte, 1990*). Indeed, the *endothelium* plays a role in the regulation of vascular tone by the release of relaxing and contracting factors, but alterations in vascular smooth muscle cell function may also contribute. Due to its anatomical position between the circulating blood and vascular smooth muscle cells, the endothelium is a primary regulator of platelet-vessel wall interaction. Finally, the endothelium and vascular smooth muscle cells take part in the regulation of proliferative responses within the blood vessel wall.

In the treatment of coronary artery disease, surgical therapy, using *arterial or venous bypass graft*, plays a crucial role. Although blood flow to the myocardium can be restored, usually using this technique similar changes as those occurring in native coronary artery disease also occur in bypass vessels, in particular, in saphenous vein grafts, thereby endangering the success of this intervention.

This review summarizes the current knowledge on endothelial dysfunction in coronary artery disease and coronary bypass graft disease.

NITRIC OXIDE AS A REGULATOR OF VASCULAR TONE

Endothelium-derived Nitric Oxide (EDNO; Fig. 1)

Endothelium-dependent relaxations can be elicited by (*Furchgott and Zawadzki,*

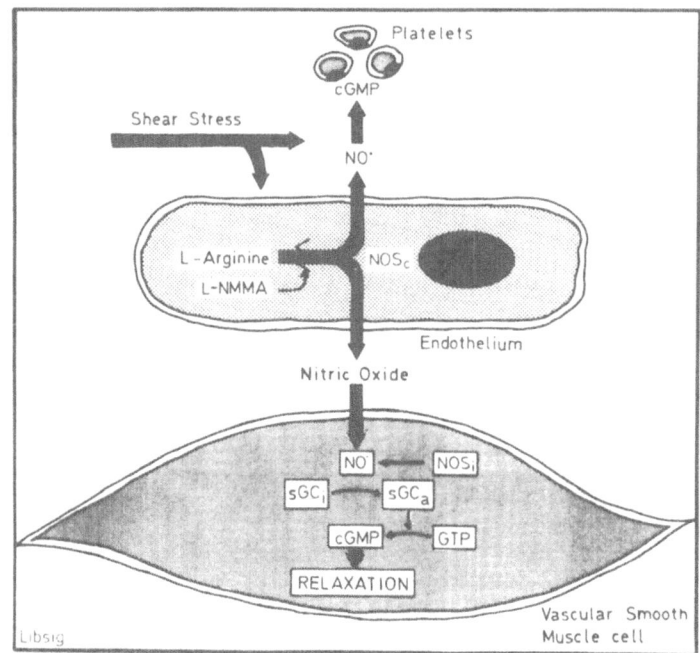

FIGURE 1: The L-arginine pathway in the blood vessel wall: Endothelial cells form nitric oxide (NO) from L-arginine via the activity of the constitutive nitric oxide synthase (NOS_C) which can be inhibited by analogues of the amino acid L-arginine, such as LN^G-monomethyl arginine (L-NMMA). NO activates soluble guanylyl cyclase (sGC) in vascular smooth muscle and platelets and causes increases in cyclic $3'5'$-guanosine monophosphate (cGMP) which mediates relaxation and platelet inhibition, respectively. Shear stress and receptor-operated agonists (not shown) stimulate the release of NO. In addition, vascular smooth muscle cells can form nitric oxide via the activity of an inducible (by tumor necrosis factor, interleukin 1 and lipopolysaccharide) form of nitric oxide synthase (NOS_i).

1992; Yang, 1991; Linder et al., 1990) neurotransmitters, hormones, substances derived from platelets and the coagulation system (*Lüscher and Vanhoutte, 1990; Furchgott and Zawadzi, 1980; Yang, 1991; Linder et al., 1990*). Physical stimuli, such as shear stress exerted by the circulating blood, induce flow-induced endothelium dependent vasodilation (*Rubanyi et al., 1986*). The relaxations are mediated by nitric oxide (NO; *Rubanyi et al., 1986; Palmer et al., 1987*). EDNO is formed from L-arginine by oxidation of the guanidine-nitrogen terminal of L-arginine (Fig. 1; *Palmer et al., 1988*). NO synthase has recently been cloned (*Bredt et al., 1991*); it is a primarily cytosolic enzyme requiring calmodulin, Ca^2+ and NADPH and has similarities with cytochrome P450 enzymes (*Bredt et al., 1990; Bredt and Snyder, 1990*). Several isoforms of the enzyme not only occur in endothelial cells, but also platelets, macrophages, vascular smooth muscle cells and in the brain (*Radomski et al., 1990; Hibbs et al., 1988;* see section 2.2). In porcine and human coronary arteries, endothelium-dependent relaxations to serotonin are inhibited by analogues of L-arginine, such as L-NG monomethyl arginine (L-NMMA), and are restored by L- but not D-arginine (*Yang et al., 1991; Tanner et al., 1990*). In quiescent arteries, L-NMMA causes endothelium - dependent contractions (Fig. 2; *Yang et al., 1991*). When infused in rabbits, L-NMMA causes long-lasting increases in blood pressure which are reversed by L-arginine (*Rees et al., 1989*). This demonstrates that the vasculature, including the coronary circulation, is in a constant state of vasodilation due to the continuous (basal and stimulated) release of NO from the endothelium.

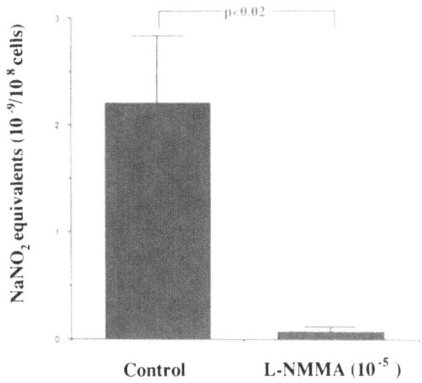

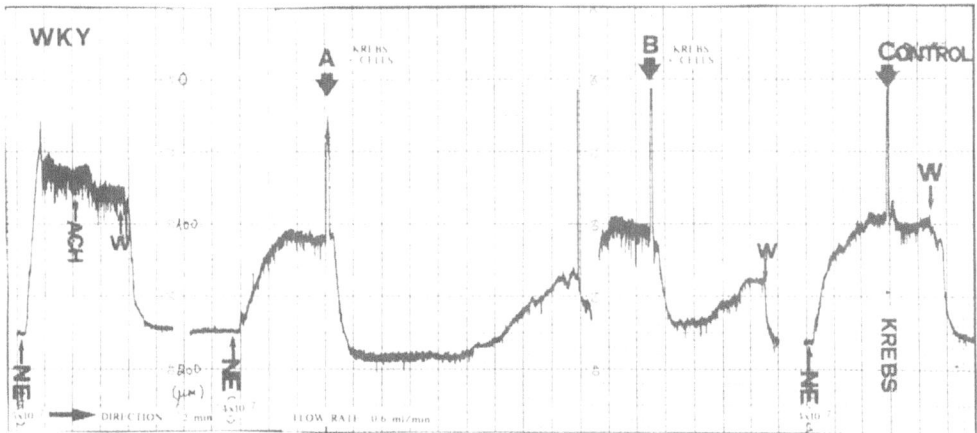

FIGURE 2: Nitric oxide release from human vascular smooth muscle cells in culture: The cells release significant amounts of nitric oxide, as measured by chemiluminescence, which can be prevented by incubation of the cells with **L-NMMA** (L-N[G]-monomethyl arginine, an inhibitor of NO formation), indicating that it is derived from L-arginine (top panel). If the medium is transferred into an arteriography system and injected into a perfused mesenteric resistance artery of the rat, biological activity can be demonstrated (bottom): First the artery is contracted with norepinephrine (NE); acetylcholine (Ach) is added to demonstrate the absence of the endothelium. After that, the agonists are washed out (w). Later, the vessel is again contracted with NE and at point A medium (Krebs + cells) incubated with the cells is injected, and a full relaxation is obtained. The flow-through with half a concentration of NO as compared to A also exhibited a relaxation with a smaller potency. After washout (w), the absence of an effect of Krebs alone is demonstrated (control). (*From Bernhart et al., 1991*).

Relaxations to EDNO are associated with an increase in cyclic 3´5´-guanosine monophosphate (cGMP) in vascular smooth muscle (Fig. 1; *Rapoport and Murad, 1983*). The inhibitor of soluble guanylyl cyclase, methylene blue, prevents the production of cGMP and inhibits endothelium-dependent relaxations (*Lüscher and Vanhoutte; Rapoport and Murad, 1983*). Thus, EDNO causes relaxations by stimulating the enzyme and in turn the formation of cGMP. Soluble guanylyl cyclase is also present in platelets and activated by EDNO (Fig. 1; *Busse et al., 1987*). Increased levels of cGMP in platelets are associated with a reduced adhesion and aggregation. Therefore, EDNO causes both vasodilatation and platelet deactivation and, thereby, represents an important antithrombotic feature of the endothelium.

Vascular Smooth Muscle-derived Nitric Oxide

Although the media of the blood vessel wall normally does not produce NO,

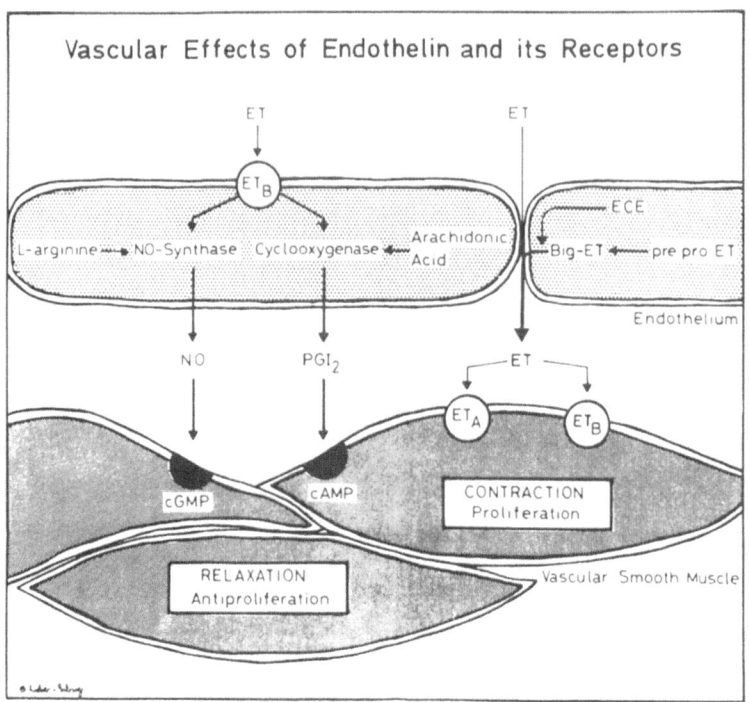

FIGURE 3: Vascular effects of endothelin and its receptors: Endothelin (ET) can activate endothelial receptors (ET_B-receptors) in the endothelial cell membrane which are linked to the production of nitric oxide (NO) and prostacyclin PGI_2). NO exerts relaxation and possibly antiproliferation by activating cGMP, while PGI_2 stimulates the formation of cAMP. Furthermore, ET can activate vascular smooth muscle cell receptors (ET_A and ET_B receptors) which mediate profound contraction and proliferation under certain conditions. (ECE: Endothelin converting enzyme).

vascular smooth muscle cells, including those obtained from human vessels, do so under culture conditions (Fig. 2; *Bernhard et al., 1991*). Indeed, endotoxin and interleukin-1 markedly stimulate the production of NO in smooth muscle (*Fleming et al., 1990; Rees et al., 1990; Schini et al., 1991*). Thus, at least two enzymes for the production of NO exist: (a) the constitutive endothelial enzyme (which is Ca^{2+}-dependent and produces picomoles of NO) and (b) the inducible form (which is Ca^{2+}-independent and produces nanomoles of NO) and is primarily expressed in smooth muscle and monocytes. Activation of the L-arginine pathway in smooth muscle cells by endotoxin, tumor necrosis factor and interleukin-1 may play a role in septic shock and explain why the cardiovascular system becomes resistant to catecholamines under these conditions. L-NMMA may provide a therapeutic tool as it is able to prevent NO formation in vascular smooth muscle stimulated with endotoxin. Preliminary data in patients in septic shock suggest that L-NMMA or a similar pharmacological tool preventing NO formation may be beneficial in patients as well (*Petros et al. 1991*).

Other Endothelium-Derived Relaxing Factors

Other endothelium-derived relaxing factors include prostacyclin (*Moncada and Vane, 1979*) and a putative endothelium-derived hyperpolarizing factor (*Lüscher and Vanhoutte, 1990; Rubanyi et al., 1986; Fletou and Vanhoutte, 1988; Komori et al., 1988; Tare et al., 1990; Standen et al., 1989*).

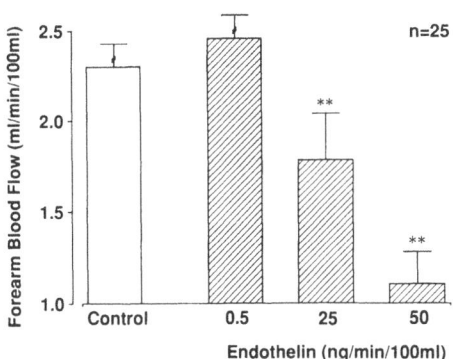

FIGURE 4: Effects of intraarterially infused endothelin-1 in the forearm circulation of normal subjects: Forearm blood flow under control conditions slightly increases at the low concentration of the peptide, but then is significantly decreased at higher concentrations (*from Kiowski et al., 1991* by permission of the American Heart Association).

ENDOTHELIUM-DERIVED CONTRACTING FACTORS (EDCF)

Cyclooxygenase-Dependent Endothelium-Derived Contracting Factor (EDCF)

Exogenous arachidonic acid evokes endothelium-dependent contractions prevented by indomethacin (an inhibitor of cyclooxygenase; *Miller and Vanhoutte, 1985*). In the human saphenous vein, acetylcholine and histamine cause endothelium--dependent contractions; in the presence of indomethacin, however, endothelium-dependent relaxations are unmasked (*Yang et al., 1991*). The products of cyclooxygenase mediating the contractions are thromboxane A_2 in the case of acetylcholine and endoperoxides (prostaglandin H_2) in that of histamine (*Yang et al., 1991*). Thromboxane A_2 endoperoxide activate both vascular smooth muscle and platelets.

Furthermore, the cyclooxygenase pathway is a source of superoxide anions which can mediate endothelium-dependent contractions either by the breakdown of NO or direct effects on vascular smooth muscle (*Katusic and Vanhoutte, 1989*). Thus, the cyclooxygenase pathway produces a variety of endothelium-derived contracting factors; their release appears particularly prominent in veins and in the cerebral and ophthalmic circulation (*Lüscher and Vanhoutte, 1990*).

Endothelin

Endothelial cells produce the 21 amino acid peptide endothelin (Fig. 3; *Lüscher and Vanhoutte, 1990; Yanagisawa et al., 1988*). Among the three peptides endothelin-1, endothelin-2 and endothelin-3, endothelial cells appear to produce, exclusively, endothelin-1 (*Lüscher and Vanhoutte, 1990*).

Translation of messenger RNA generates preproendothelin, which is converted to big endothelin (*Yanagisawa et al., 1988*); its conversion to endothelin-1 by the endothelin converting enzyme is necessary for the development of full vascular activity (*Kimura et al., 1988*). The expression of messenger RNA and the release of the peptide is stimulated by thrombin, transforming growth factor-beta, interleukin-1, epinephrine, angiotensin, arginine vasopressin, calcium ionophore and phorbol ester (*Lüscher and Vanhoutte, 1990; Yanagisawa et al., 1988; Kimura et al., 1988; Boulanger and Lüscher, 1990; Lüscher et al., 1992*).

Endothelin-1 is a potent vasoconstrictor, both *in vitro* and *in vivo* (Fig. 4; *Lüscher et al., 1992*). In the coronary circulation and the human forearm, endothelin causes vasodilation at lower and marked contractions at higher concentrations (*Kiowski et al., 1991*), which, in the heart, eventually leads to ischemia, arrhythmias and death.

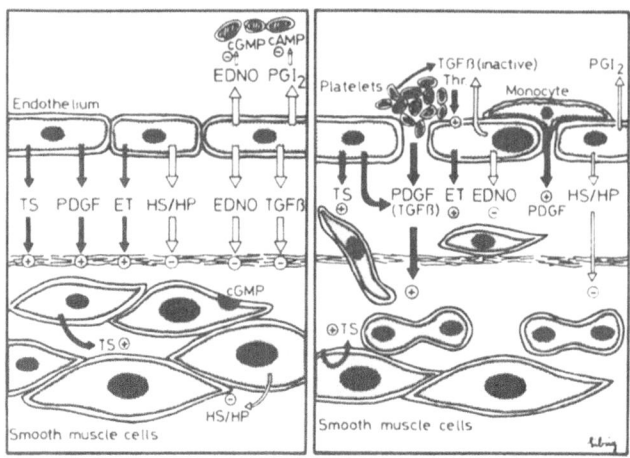

FIGURE 5: Endothelium-dependent control mechanisms of vascular growth: Under normal conditions, the vascular wall remains quiescent (left panel). This may be related to the predominance of inhibitory (white arrows; ⊖) stimuli, such as heparin, heparin sulfates (HS/HP), endothelium-derived nitric oxide (EDNO) and transforming growth factorβ1 (TGF$_{\beta 1}$). EDNO may exert antiproliferative effects, either directly on vascular smooth muscle cells via activating 3´5´-cyclic guanosine monophosphate (cGMP) or by inhibiting platelet function (together with prostacyclin; PGI$_2$), which then reduces the local concentrations of platelet-derived growth factor (PDGF). In contrast, in diseased arteries (right panel), platelets and monocytes adhere and release potent growth factors, in particular PDGF. TS = thrombospondin; ET = endothelin.

Also, in human arterial and venous coronary bypass vessels, endothelin causes marked contractions (*Lüscher et al., 1990*).

The circulating levels of endothelin-1 are low indicating a low production rate under physiological conditions (*Lüscher et al, 1992; Suzuki et al., 1989*). This appears to be related to the presence of potent inhibitory mechanisms and preferential release toward smooth muscle cells (*Yoshimoto et al., 1990*). Indeed, three inhibitory mechanisms regulating endothelin production have been delineated: (1) cGMP-dependent inhibition (*Boulanger and Lüscher, 1990*); (2) cAMP-dependent inhibition (*Yokokawa et al., 1991*) and (3) an inhibitory factor produced by vascular smooth muscle cells (*Stewart et al., 1990*). The cGMP-dependent mechanism can be activated by EDNO, nitroglycerine, 3- morpholino sydnominine (SIN-1; *Boulanger and Lüscher, 1991*) and atrial natriuretic peptide, which activates particulate guanylyl cyclase (*Saijounaa et al., 1990*). Thus, after inhibition of the endothelial L-arginine pathway, the thrombin-induced production of endothelin is augmented (*Lüscher et al., 1992*); on the other hand, SIN-1 prevents the thrombin-induced endothelin release via a cyclic GMP-dependent mechanism (*Boulanger and Lüscher, 1991*). Endothelin can also release NO and prostacyclin from endothelial cells which may represent a negative feedback mechanism (*Warner et al.,; Dohi and Lüscher, 1991*).

The contraction to endothelin is not related to direct activation of voltage-operated Ca^{2+} channels on smooth muscle, since Ca^{2+} antagonists do not prevent its binding nor its effects in most blood vessels (*Clozel et al., 1989; Yang et al., 1990*). However, the peptide indirectly activates voltage-operated Ca^{2+} channels in the porcine coronary artery where Ca^{2+} antagonists attenuate endothelin-induced vasoconstriction (*Goto et al., 1989*). In the human forearm circulation, endothelin-1 induces potent contractions which are prevented by nifedipine and verapamil, unmasking the vasodilator effects of the peptide (*Kiowski et al., 1991*). The vasodilator effects of endothelin are related to the endothelial production of prostacyclin (*Dohi and Lüscher, 1991*) although NO may contribute (*Warner et al., 1989*).

60

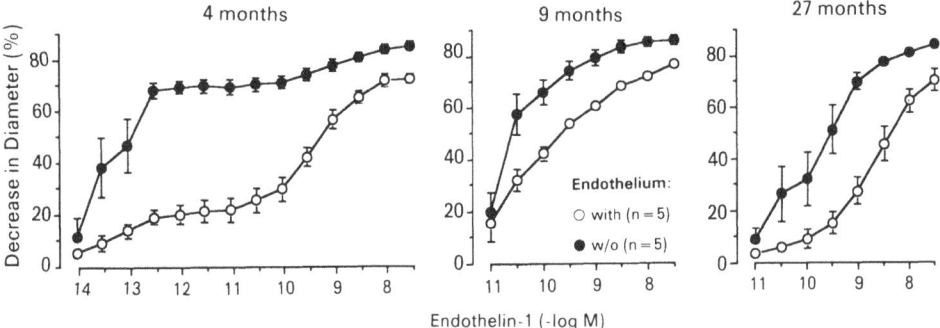

FIGURE 6: Effects of aging on the response of perfused mesenteric resistance arteries of Fischer rats: Preparations without endothelium (●) are extremely sensitive to the peptide in young animals, while aging is associated with a desensitization to the peptide. Similarly, the endothelium markedly reduces the effects of endothelin (0), particularly in young animals. This inhibitory effect of the endothelium markedly decreases with age (*from Dohi and Lüscher, 1990*).

EDNO also interacts with endothelin at the level of vascular smooth muscle. The contractions to endothelin are enhanced after endothelial removal indicating that basal production of EDNO reduces its response (*Lüscher et al., 1990*). Stimulation of the formation of EDNO by acetylcholine reverses endothelin-induced contractions in most blood vessels although this mechanism appears to be less potent in veins (*Lüscher et al., 1990*).

THE ENDOTHELIUM AND REGULATION OF VASCULAR GROWTH

Removal of the endothelium is a procedure which invariably leads to a proliferative response in the blood vessel wall with intimal hyperplasia. Important growth inhibitors produced by the endothelium are heparin and heparin sulfates and most likely also transforming growth factor β (Fig. 5; *Hannan et al., 1988; Battegay et al., 1990; Yang et al., 1992*). On the other hand, endothelial cells can produce growth factors, such as basic fibroblast growth factor, platelet-derived growth factor and possibly also endothelin that contribute to proliferative responses, at least under certain conditions (*Hannan et al., 1988; Yang et al., 1992; Dubin et al., 1989*). Not much is known yet about the regulation of endothelium-dependent mechanisms regulating antiproliferation and proliferation, but it has to be assumed that under normal conditions, inhibitory stimuli prevail.

RISK FACTORS FOR CORONARY ARTERY DISEASE

Aging

Aging is one of the most important determinants of vascular disease. In the rat, aging leads to increased formation of the cyclooxygenase-dependent endothelium--derived contracting factor (prostaglandin H_2; *Koga et al., 1989*) as well as a mild decrease in the release of EDRF (*Dohi and Lüscher, 1990*). In contrast, the responsiveness of smooth muscle to NO-forming compounds does not change with aging, while that to endothelin decreases (Fig. 6, *Dohi and Lüscher, 1990*). In the human coronary microcirculation, the increase in coronary flow induced by intraarterial infusions of acetylcholine declines with aging (*Vital et al., 1990*).

Regenerated Endothelium

After mechanical denudation of the porcine coronary artery, regenerated

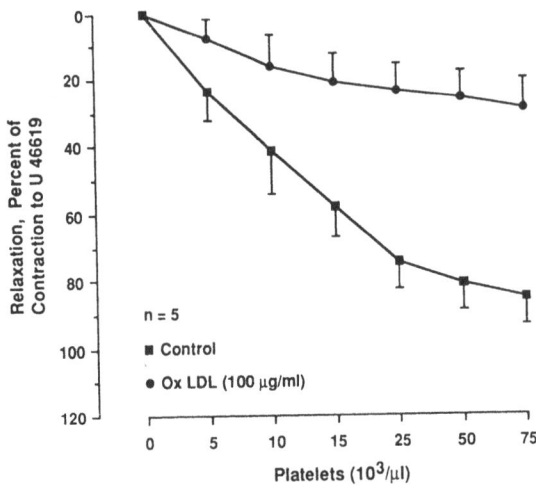

FIGURE 7: Effects of aggregating platelets in the porcine coronary artery with endothelium: Increasing concentrations of platelets cause endothelium-dependent relaxations (in the presence of ketanserin) averaging almost 90% of the precontraction to the thromboxane analog U 4669. This relaxation (which is mediated by nitric oxide; not shown) is markedly inhibited in the presence of oxidized low-density lipoproteins (ox-LDL) (*from Tanner et al., 1991 by permission of the American Heart Association*).

endothelial cells have an impaired capacity to release EDNO in response to platelet-derived serotonin because of a defect of the Gi-protein linked to the endothelial 5HT$_1$-serotonergic receptor (*Shimokawa et al., 1989*). As similar changes must occur during life-time, these functional alterations may contribute to age-dependent impairments of endothelial function in the coronary circulation.

In addition, balloon injury induces nitric oxide synthase activity in rat carotid artery (*Joly, 1992*). The formation of endothelin under these conditions is unknown at this point. It is possible, however, that these functional alterations of the endothelium also play a role after percutaneous transluminal coronary angioplasty. Dysfunction of the endothelium, particularly in response to platelet-derived products, may increase platelet adhesion and thereby provide high local concentrations of platelet-derived growth factor which may contribute to proliferative responses at sites of endothelial dysfunction.

Lipoproteins and Hypercholesterolemia

Morphologically, the endothelium remains intact in early stages of atherogenesis. Functionally, however, pronounced alterations occur (*Lüscher and Vanhoutte, 1990*). Particularly, oxidized low density lipoproteins (OX-LDL) are present in human atherosclerotic lesions (*Ylä-Harttuala et al., 1989*). In the porcine coronary artery, OX-LDL inhibit endothelium-dependent relaxations to platelets, serotonin and thrombin (Fig. 7; *Tanner et al., 1991*). In contrast, relaxations to the NO-donor SIN-1 are well maintained excluding a reduced responsiveness of vascular smooth muscle to EDNO. The inhibition is specific for OX-LDL, as it is not induced by comparable concentrations of native LDL. In the rabbit aorta, the effect of OX-LDL is mimicked by lysolecithin (a characteristic component of OX-LDL; *Kugiyama et al., 1990*).

OX-LDL appears to activate an endothelial receptor distinct from the LDL receptor such as the scavenger receptor (*Tanner et al., 1991a; Tanner et al., 1991b*); indeed, dextran sulfate, a competitive antagonist of modified LDL at this receptor (*Kugiyama et al., 1990*), prevents the endothelial effects of OX-LDL. The inhibitor of NO production L-NMMA exerts a similar effect, suggesting that OX-LDL specifically

62

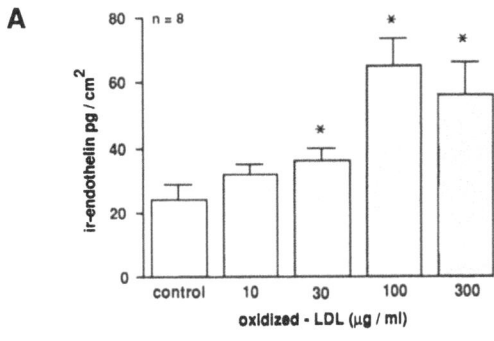

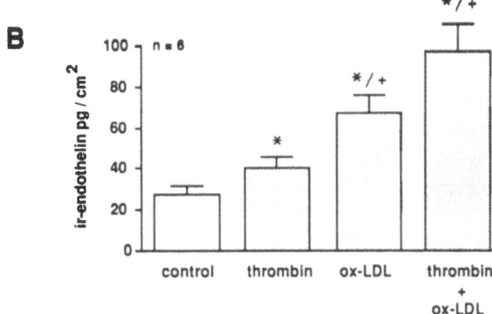

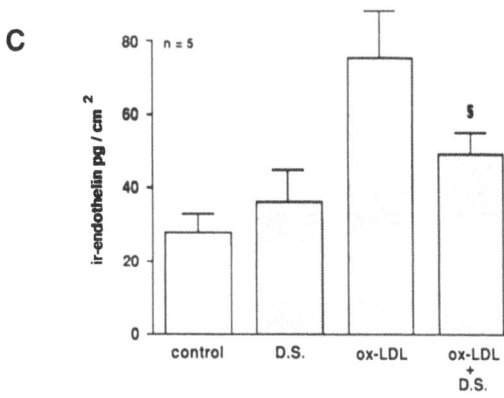

FIGURE 8: Oxidized low density lipoproteins (LDL) induce the release of endothelin from intact porcine aorta: As compared to control conditions, the lipoproteins concentration-dependently increase the release of immunoreactive (ir) endothelin (panel A). The effects of ox-LDL are additive to those of thrombin (panel B). Finally, the putative scavenger receptor antagonist dextran sulfate (D.S.) does not affect basal formation of ir-endothelin (control) but markedly inhibits those elicited by ox-LDL, indicating that it is mediated by activation of the scavenger receptor (panel C).

$^{*}=p<0.05$ vs control. (*from Boulanger, et al., 1990, by permission of the American Heart Association*).

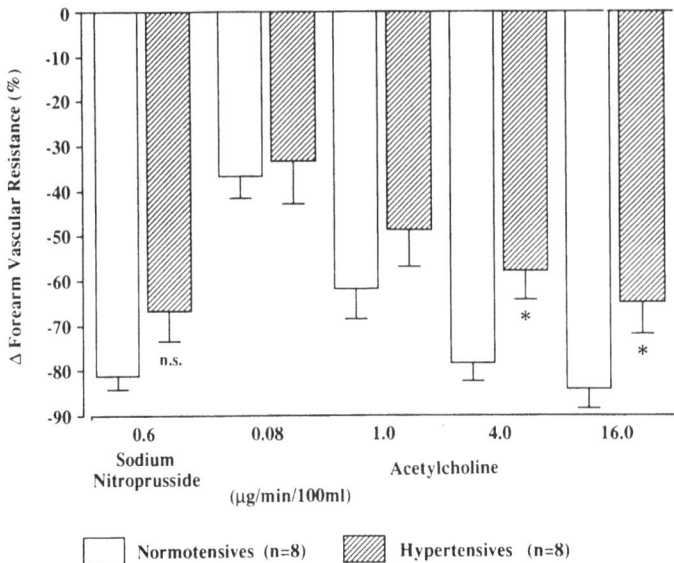

FIGURE 9: Effects or intraarterial infusion of acetylcholine in the forearm circulation of normotensive and hypertensive subjects: The decrease in forearm vascular resistance is less pronounced in patients with essential hypertension. (*from Linder et al., 1990, by permission of the American Heart Association*).

interferes with the L-arginine pathway. NO synthetase, however, remains unaffected as L-arginine evokes a full relaxation in vessels treated with OX-LDL. Pretreatment with L-arginine restores the response to serotonin in vessels treated with OX-LDL. Thus, OX-LDL may interact with the intracellular availability of L-arginine (*Tanner et al., 1991a; Tanner et al., 1991b*). This mechanism may also occur *in vivo*, as in hypercholesterolemic pigs a similar inhibition of endothelium-dependent relaxation to serotonin occurs, as in coronary arteries exposed to OX-LDL (*Shimokawa and Vanhoutte, 1989*). In humans with hypercholesterolemia, L-arginine infusion augments the blunted increase in coronary blood flow in response to acetylcholine (*Drexler et al., 1991*); in contrast the loss of endothelium-dependent vasodilation to acetylcholine in epicardial coronary arteries is unaffected by the amino acid, possibly because of the presence of fully developed atherosclerosis.

In addition to their effect on the L-arginine pathway, both native and OX-LDL inactivate NO (*Galle et al., 1991*). OX-LDL not only inhibit endothelium-dependent relaxation but also cause endothelium-dependent contraction (*Simon et al., 1990*). In the rabbit femoral artery, OX-LDL potentiate contractions to KCl as well as to receptor-operate vasoconstrictors (*Galle et al., 1990*).

OX-LDL also induce the expression of messenger RNA for endothelin in cultured aortic endothelial cells as well as the release of the peptide from the intact porcine aorta (Fig. 8; *Boulanger et al., 1990*). As threshold and low concentrations of endothelin, which by themselves evoke no appreciable vascular effect, potentiate contractions induced by serotonin in the human coronary artery and to norepinephrine and serotonin in the human internal mammary artery, even small increases in endothelin production may be relevant (*Yang et al., 1990*).

Hypertension

Endothelium-dependent relaxations to acetylcholine are reduced in hypertensive rats (*Lüscher, 1990; Dohi et al., 1990*). Similarly, the vasodilator effects of acetylcholine in the human forearm of hypertensive subjects is attenuated (Fig. 9; *Linder et al., 1990*). In the spontaneously hypertensive rat, the reduced response to acetylcholine is related to the production of a cyclooxygenase-dependent endothelium-derived

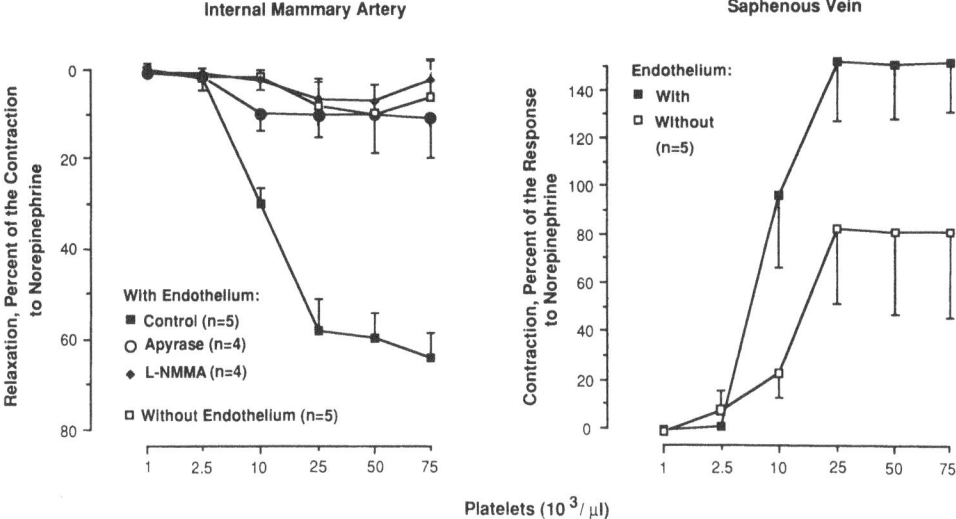

FIGURE 10: Effects of aggregating platelets in the human internal mammary artery (left panel) and saphenous vein (right panel): While in the mammary artery, platelets evoke potent endothelium-dependent relaxations mediated by platelet-derived adenosine diphosphate (ADP; which is broken down by the enzyme apyrase) and in turn the release of endothelium-derived nitric oxide (inhibited by LN^G-mono-methyl arginine; L-NMMA), aggregating platelets evoke potent contractions in the saphenous vein (*modified from Yang et al., 1991, by permission*).

contracting factor (i.e. prostaglandin H_2), while in other forms of hypertension of the rat, a reduced formation of EDNO predominates (*Lüscher, 1990*). In the mesenteric microcirculation, intraluminal activation of the endothelium is dysfunctional indicating a predominant alteration of that surface of the endothelium which is most exposed to high blood pressure (*Lüscher, 1990*).

In contrast, the coronary circulation - at least as judged from the spontaneously hypertensive rat - appears less prone to hypertensive endothelial dysfunction. Indeed, in epicardial coronary arteries of the spontaneously hypertensive rat, aging and hypertension only minimally reduce the response to acetylcholine (*Tschudi et al., 1991*). This suggests that, in the absence of hyperlipidemia, hypertension exerts only a mild effect on the coronary endothelium.

Diabetes

Insulin and glucose both interfere with the release of EDNO. Indeed, in normal human coronary arteries, insulin causes endothelium-dependent relaxations (*Thom et al., 1988*). Increasing concentrations of glucose, on the other hand, reduce the release of EDNO (*Bucala et al., 1991*). In addition, in the aorta of the rabbit, diabetes is associated with an increased formation of endothelium-derived thromboxane A_2 which inhibits the effects of EDNO (*Lüscher and Vanhoutte, 1990; Tesfamariam et al., 1989*). In the *corpora cavernosa* of diabetic patients, the vasodilator effects of acetylcholine are reduced, while those to sodium nitroprusside are maintained (*De Tejada et al., 1989*). Hence, endothelial dysfunction does occur in the microcirculation of diabetic patients *in vivo* and may contribute to tissue ischemia and its complications.

ATHEROSCLEROSIS

Atherosclerosis is the final common pathway of vascular disease induced by most cardiovascular risk factors. Morphologically, it is characterized by an increased platelet-vessel wall interaction (i.e. platelets, monocytes), endothelial cell polymor-

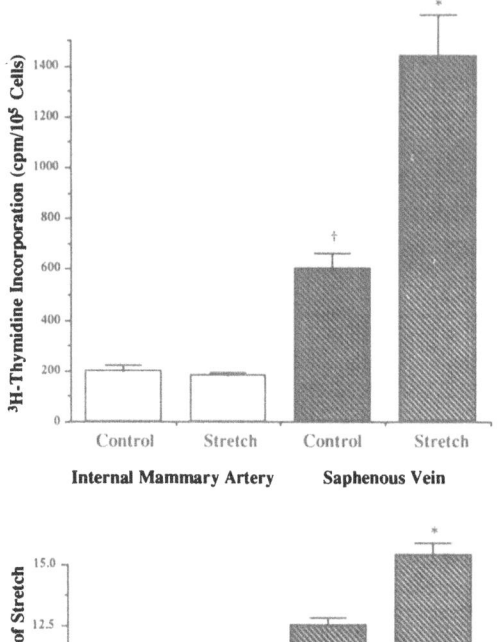

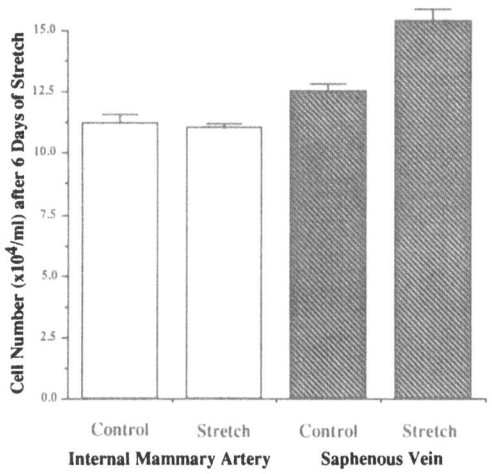

FIGURE 11: Effects of pulsatile stretch on proliferation of vascular smooth muscle cells obtained from the human internal mammary artery (□) or saphenous vein (■) ³H-thymidine incorporation is markedly increased in the saphenous vein in response to pulsatile stretch (top) as is cell number (bottom). (*from Predel et al., 1992, by permission*).

phism, proliferative changes of vascular smooth muscle and fibroblasts, and at late stages, also, endothelial denudation, cap rupture and thrombosis (*Lüscher and Vanhoutte, 1990*).

In porcine coronary arteries, established atherosclerosis severely impairs endothelium-dependent relaxations to serotonin and also reduces endothelium-dependent relaxations to bradykinin, which are maintained in hypercholesterolemia (*Shimokawa and Vanhoutte, 1989*). However, endothelium-independent relaxations to nitrovasodilators remain preserved except in severely atherosclerotic arteries (*Lüscher and Vanhoutte, 1990*). In atherosclerotic human coronary arteries, endothelium-dependent relaxations to substance P, bradykinin, aggregating platelets and calcium ionophore are attenuated (*Bossaller et al., 1987; Förstermann et al., 1988*) and *in vivo* acetylcholine as well as serotonin cause paradoxical vasoconstriction (*Ludmer et al., 1986; Golino et al., 1991*).

Controversy exists as to the mechanism responsible for the marked impairment or loss of endothelium-dependent relaxations in atherosclerosis. The bioassayable **EDRF** release in porcine coronary artery with hypercholesterolemia and atherosclerosis clearly is reduced (*Shimokawa and Vanhoutte, 1989*). Direct measurements of nitric oxide in the rabbit aorta, however, suggest an increased formation of NO with a concomitant massive breakdown of the endogenous nitrovasodilator (*Minor et al., 1990*). The latter observation would suggest an increased formation of superoxide radicals and other products inactivating NO and/or decreased activity of superoxide dismutase in the blood vessel wall in atherosclerosis. Furthermore, it is conceivable, that atherosclerosis in the more developed stages, with marked invasion of monocytes and other blood cells, induces nitric oxide synthase in the subintimal space and vascular smooth muscle cells.

CORONARY BYPASS GRAFT FUNCTION AND PATENCY

In patients with coronary artery disease, surgical therapy involves implantation of an arterial or venous bypass graft using the internal mammary, gastroepiploic artery and/or saphenous vein (*Loop et al., 1986*). The mammary artery has a remarkably higher patency rate than the saphenous vein (*Loop et al., 1986*).

Endothelium-derived vasoactive factors may importantly contribute to graft function as they determine the antithrombotic properties and the regulation of blood flow. In addition, the factors may have antiproliferative and proliferative properties determining the late changes occurring in coronary bypass grafts (*Lüscher et al., 1988*).

Endothelial Function

The mammary artery exhibits much more pronounced endothelium-dependent relaxations as compared to the saphenous vein, because the release of EDNO by receptor operated agonists and in particular by aggregating platelets is more efficient in the artery than the vein (Fig. 10; *Lüscher et al., 1988; Yang et al., 1991*). Particularly, the release of EDNO in response to platelet-derived adenosine diphosphate represents an important antithrombotic property (Fig. 1; *Lüscher and Vanhoutte, 1990; Yang et al., 1992*). The gastroepiploic artery releases comparable amounts of EDNO as the mammary artery but exhibits more pronounced contractions (presumably because it represents a muscular rather than elastic artery; *Yang, 1992*).

Vascular Smooth Muscle

In particular, the intimal hyperplasia and stenosis formation in venous grafts is related morphologically to vascular smooth muscle cell proliferation. Important potential factors in that process are hemodynamic changes after implantation in the coronary circulation (i.e. change from laminar flow in the venous circulation to pulsatile flow in the coronary circulation), as well as growth factors, most importantly, platelet-derived growth factor, which may be present in high concentrations, particularly at sites of endothelial denudation during surgical preparation. Interestingly, if vascular smooth muscle cells obtained from the human internal mammary artery and saphenous vein are cultured and exposed to pulsatile stretch in a specially designed computer-guided system (Flexcell system), vascular smooth muscle cells from the vein exhibit marked proliferative responses both in terms of ^{13}H-thymidine incorporation, as well as cell number, while the internal mammary artery does not exhibit proliferative responses to that stimulus (Fig. 11; *Predel et al., 1992*). In addition, platelet-derived growth factor is a potent mitogen for saphenous vein vascular smooth muscle but has very little effect in the human internal mammary artery vascular smooth muscle cells or low fetal calf serum does exhibit proliferative responses in that tissue (*Yang et al., 1992*).

These differences in endothelial and vascular smooth muscle function of various bypass graft vessels may play an important role in graft function and patency and in turn for the survival of patients undergoing coronary bypass surgery.

ACKNOWLEDGMENTS

The authors are indebted to Amanda de Sola Pinto, Sabine Bohnert and to Bernadette Weber-Libsig for their help in the preparation of the manuscript. Original research reported in the manuscript was supported by grants of the Swiss National Research Foundation (No. 32-32541.91 and SCORE Grant No. 3231-025.150), the Helmut Horten Foundation and the Stanley Thomas Johnson Foundation.

REFERENCES

Battegay, E.J., Raines, E.W., Seifert, R.A., Bowen-Pope, D.F., Ross, R., 1990, TGF-p induces bimodal proliferation of connective tissue cells via complex control of an autocrine PDGF loop. *Cell* 63:515.

Bernhardt, J., Tschudi M.R., Dohi Y., Gut I., Urwyler B., Bühler F.R., Lüscher T.F., 1991, Release of nitric oxide from human vascular smooth muscle cells. *Biochem. Biophys. Res. Comm.* 180 (2):907.

Bossaller, C., Habib, G.B., Yamamoto, H. Williams, C., Wells, S., Henry, P.D., 1987, Impaired muscarinic endothelium relaxation and cyclic guanosine 3´5´-monophosphate formation in atherosclerotic human coronary artery and rabbit aorta. *J. Clin. Invest.* 79: 170.

Boulanger, C.M., Lüscher, T.F., 1991, Hirudin and nitric oxide donors inhibit the thrombin-induced release of endothelin from the intact porcine aorta. *Circ. Res.* 68:1768.

Boulanger, C., Lüscher, T.F., 1990, Release of endothelin from the porcine aorta: Inhibition by endothelium-derived nitric oxide. *J. Clin. Invest.* 85:587.

Boulanger, C.V., Tanner, F.C., Hahn, A.W.A., Werner, A., Lüscher, T.F., 1992, Oxidized low-density lipoproteins mRNA expression and release of endothelin from human and porcine endothelium *Circ. Res.* 70:1991.

Bredt, D.S., Snyder, S.H., 1990, Isolation of nitric oxide synthetase, a calmodulin-requiring enzyme. *Proc. Natl. Acad. Sci. USA* 87:682.

Bredt, D.S., Hwang, P.M., Glatt, C.E., Lowenstein, C., Reed, R.R., Snyder S.H., 1991, Cloned and expressed nitric oxide synthase structurally resembles cytochrome P-450 reductase. *Nature* 351:714.

Bucala, R., Tracey, K.J., Cerami, A., 1991, Advanced glycosylation products quench nitric oxide and mediate defective endothelium-dependent vasodilation in experimental diabetes. *J. Clin. Invest.* 87 (22):432.

Busse, R., Lückhoff, A., Bassenge, E., 1987, Endothelium-derived relaxant factor inhibits platelet activation. *Naunyn-Schmiedeberg's Arch. Pharmacol.* 336:566.

Clozel, M., Fischli, W., Guilly, C., 1989, Specific binding of endothelin on human vascular smooth muscle cells in culture. *J. Clin. Invest.* 83:1758.

De Tejada, I.D., *et al.*, 1989, Impaired neurogenic and endothelium-mediated relaxation of penile smooth muscle from diabetic men with impotence. *N. Engl. J. Med. 320:1025.*

DiCorleto, P.E., Fox, P.L., 1990, Growth factor production by endothelial cells. In: Endothelial cells. U. Ryan (ed), Boca Raton, FL, *CRC Press,* II:51-62.

Dohi, Y., Lüscher, T.F., 1991, Endothelin-1 in hypertensive resistance arteries:intraluminal and extraluminal dysfunction. *Hypertension* 18:543.

Dohi, Y., Lüscher, T.F., 1990, Aging differentially affects direct and indirect actions of endothelin-1 in perfused mesenteric arteries of the rat. *Br. J. Pharmacol.* 100:889.

Dohi, Y., Thiel, M.A., Bühler, F.R., Lüscher, T.F., 1990, Activation of endothelial L-arginine pathway in resistance arteries. *Hypertension.* 15:170.

Drexler, H., Zeiher, A.M., Meinzer, K., Just, H., 1991, Correction of endothelial dysfunction in coronary microcirculation of hypercholesterolemic patients by L-arginine. *Lancet* 338:1546.

Dubin, D., Pratt, R.E., Cooke, J.P., Dzau, V.J., 1989, Endothelin, a potent vasoconstrictor, is a vascular smooth muscle mitogen. *J. Vasc. Med. Biol.* 1:13.

Feletou, M., Vanhoutte, P.M., 1988, Endothelium-dependent hyperpolarization canine coronary smooth muscle. *Br. J. Pharmacol.* 93:515.

Fleming, I., *et al.*, 1990, Incubation with endotoxin activates the L-arginine pathway in vascular tissue. *Biochem. Biophys. Res. Commun.* 171 (2):562.

Föstermann, U., *et al.*, 1988, Selective attenuation of endothelium-mediated vasodilation in atherosclerotic human an coronary arteries. *Circ. Res.* 62:185.

Furchgott, R.F., Zawadzki, J.V., 1980, The obligatory role of endothelial cells in the relaxation of arterial smooth muscle by acetylcholine. *Nature* 299:373.

68

Galle, J., Mülsch, A., Busse, R., Bassenge, E., 1991, Effects of native and oxidized low-density lipoproteins on formation and inactivation of endothelium-derived relaxing factor. *Arterioscl. Thromb.*
11:198.

Galle, J., Bassenge, E., Busse R., 1990, Oxidized low density lipoproteins potentiate vascoconstrictions
to various agonists by direct interaction with vascular smooth muscle. *Circ. Res.* 66:1287.

Golino, P., *et al.*, 1991, Divergent effects of serotonin on coronary-artery dimensions and blood flow in
patients with coronary atherosclerosis and control patients. *N. Engl. J. Med.* 324:641.

Goto, K., Kasuya, Y., Matsuki, N., Takuwa, Y., Kurihara, H., Ishikawa, T., Kimura, S., Yanagisawa,
M., Masaki, T., 1989, Endothelin activates the dihydropyridine-sensitive, voltage-dependent
Ca^{2+} channel in vascular smooth muscle. *Proc. Natl. Acad. Sci. USA* 86:391.

Hannan, R.L., *et al.*, 1988, Endothelial cells synthesize basic fibroblast growth factor and transforming
growth factor beta. *Growth Factors* 1:7.

Hibbs, J.B., Traintor, R.R., Vavrin, Z., Rachlin, E.M., 1988, Nitric oxide: A cytotoxic activated macrophage molecule. *Biochem. Biophys. Res. Comm.* 157(1):87.

Joly, G.A., Schini, V.B., Vanhoutte, P.M., 1992, Balloon injury and interleukin-1 beta induce nitric
oxide synthase activity in rat carotid arteries. *Circ. Res.* 71:331.

Katusic, Z.S., Vanhoutte, P.M., 1989, Superoxide anion is an endothelium-derived contracting factor.
Am. J. Physiol. 357:H33.

Kimura, S., *et al.*, 1988, Structure-activity relationships of endothelin: importance of the c-terminal
moiety. *Biochem. Biophys. Res. Comm.* 156:1182.

Kiowski, W., Lüscher, T.F., Linder, L., Bühler, F.R., 1991, Endothelin-1 induced vasoconstriction in
man: reversal by calcium channel blockade but not by nitrovasodilators or
endothelium-derived relaxing factor. *Circulation* 83:469.

Knowles, R.G., Palacios, M., Palmer, R.M.J., Moncada, S., 1989, Formation of nitric oxide from
L-arginine in the central nervous system: a transduction mechanism for stimulation of the
soluble guanylate cyclase. *Proc. Natl. Acad. Sci. USA* 86:1.

Koga, T., *et al.*, 1989, Age and hypertension promote endothelium-dependent contractions to acetylcholine in the aorta of the rat. *Hypertension* 14:542-548.

Komori, K., Lorenz, R.R., Vanhoutte, P.M., 1988, Nitric oxide, acetylcholine, and electrical and
mechanical properties of canine arterial smooth muscle. *Am. J. Physiol.* 255:H207.

Kugiyama, K., *et al.*, 1990, Impairment of endothelium-dependent arterial relaxation by lysolecithin in
modified low-density lipoproteins. *Nature* 344:160.

Linder, L., Kiowski, W., Bühler, F.R., Lüscher, T.F., 1990, Indirect evidence for release of
endothelium-derived relaxing factor in human forearm circulation *in vivo*: Blunted response in
essential hypertension. *Circulation* 81:1762.

Loop, F.D., Lytle, B.W., Cosgrove, D.M., Stewart, R.W., Goorastic, M., Williams, G.W., Golding, L.A.
R., Gill, C.G., Taylor, P.C., Sheldon, W.C., Proudfit, W.L., 1986, Influence of the internal
mammary-artery graft on 10- year survival and other cardiac events. *N. Engl. J. Med.* 314:1.

Ludmer, P.L., Selwyn, A.P., Shook, T.L., Wayne, R.R., Mudge, G.H., Alexander, R.W., Ganz, P. Paradoxical vasoconstriction induced by acetylcholine in atherosclerotic coronary arteries.

Lüscher, T.F., *et al.*, 1991, Difference between endothelium-dependent relaxations in arterial and in
venous coronary bypass grafts. *N. Engl. J. Med.* 319:462.

Lüscher, T.F., Vanhoutte, P.M., 1990, The endothelium modulator or cardiovascular function. *CRC
Press, Boca Raton, FL., U.S.A.* 1-215.

Lüscher, T.F., 1991, Vascular biology of coronary bypass grafts. *Current Opinion Cardiol.* 6:868-876.

Lüscher, T.F., Boulanger, C.M., Dohi, Y., Yang, Z., 1992, Endothelium-derived contracting factors.
Hypertension 19:117.

Lüscher, T.F., Yang, Z., Tschudi, M., von Segesser, L., Stulz P., Boulanger, C., Siebenmann, R.,
Turina, M., Bühler, F.R. *et al.*, 1990. Interaction between endothelin-1 and endothelium-
derived relaxing factor in human arteries and veins. *Circ. Res.* 66(4):1088.

Lüscher, T.F., 1990, Imbalance of endothelium-derived relaxing and contracting factor: A new concept
in hypertension? *Am. J. Hypertens.* 3:317.

Miller, V.M., Vanhoutte, P.M., 1985, Endothelium-dependent contractions to arachidonic acid are
mediated by products of cyclooxygenase in canine veins. *Am. J. Physiol.* 248:H432.

Minor, R.L. Jr, Myers, P.R., Guerra, R. Jr, Bates, J.N., Harrison D.G., 1990, Diet-induced
atherosclerosis increases the release of nitrogen oxides from rabbit aorta. *J. Clin. Invest.*
86:2109.

Moncada, S., Vane, J.R., 1979, Pharmacology and endogenous roles of prostaglandin endoperoxides,
thromboxane Az and prostacyclin. *Pharmacol. Rev.* 30:293.

Myers, P.R., *et al.*, 1990, Vasorelaxant properties of the endothelium-derived relaxing factor more
closely resemble S-nitrosocysteine than nitric oxide. *Nature* 345:161.

Palmer, R.M.J., Ashton, D.S., Moncada, S., 1988, Vascular endothelial cells synthesize nitric oxide from L-arginine. *Nature* 333:664.

Palmer, R.M.I., Ferrige, A.G., Moncada, S., 1987, Nitric oxide release accounts for the biological activity of endothelium-derived relaxing factor: *Nature* 327:524.

Petros, A., Bennett, D., Vallance, P., 1991, Effect of nitric oxide synthase inhibitors on hypotension in patients with septic shock. *Lancet* 338:1557.

Predel, H.-G., Yang, Z., von Segesser, L., Turina, M., Bühler, F.R., Lüscher, T.F., 1992, Pulsatile stretch stimulates growth of saphenous vein but not mammary artery smooth muscle: Implications for coronary bypass graft disease. *Lancet* 340:878.

Radomski, M.W., Palmer, R M.J., Moncada, S., 1990, An L-arginine/nitric oxide pathway present in human platelets regulates aggregation. *Proc. Natl. Acad. Sci. USA* 87:5193.

Rapoport, R.M., Murad, F., 1983, Agonist-induced endothelium-dependent relaxation in rat thoracic aorta may be mediated through cGMP. *Circ. Res.* 52:352.

Rees, D.D., Cellek, S., Palmer, R.M.J., Moncada, S., 1990, Dexamethasone prevents the induction by endotoxin of a nitric oxide synthase and the associated effects on vascular tone: an insight into endotoxin shock. *Biochem. Biophys. Res. Comm.* 173:541.

Rees, D.D., Palmer, R.M.J., Moncada, S., 1989, Role of endothelium-derived nitric oxide in the regulation of blood pressure. *Proc. Natl. Acad. Sci. USA* 86:3375.

Richard, V., Tanner, F.C., Tschudi, M., Lüscher, T.F., 1990, Different activation of L-arginine pathway by bradykinin, serotonin, and clonidine in coronary arteries. *Am. J. Physiol.* 259:H1433.

Rubanyi, G.M., Romero, J.C., Vanhoutte, P.M., 1986, Flow-induced release of endothelium-derived relaxing factor. *Am. J. Physiol.* 250:H1145.

Rubanyi, G.M., Vanhoutte, P.M., 1986, Superoxide anions and hyperoxia inactivate endothelium-derived relaxing factor. *Am. J. Physiol.* 250:H822.

Saijonmaa, 0., Ristimäki, A., Fyhrquist, F., 1990, Atrial natriuretic peptide, nitroglycerine, and nitroprusside reduce basal and stimulated endothelin production from cultured endothelial cells. *Biochem. Biophys. Res. Comm.* 173:514.

Schini, V.B., Junquero, D.C., Scott-Burden, T., Vanhoutte, P.M., 1991, Interleukin-1 beta induces the production of an L-arginine-derived relaxing factor from cultured smooth muscle cells from rat aorta. *Biochem. Biophys. Res. Comm.* 176(1):114.

Shimokawa, H., Flavahan, N.A., Vanhoutte, P.M., 1989, Natural course of the impairment of endothelium-dependent relaxations after balloon endothelium-removal in porcine coronary arteries. *Circ. Res.* 65:740.

Shimokawa, H., Vanhoutte, P.M., 1989, Impaired endothelium-dependent relaxation to aggregating platelets and related vasoactive substances in porcine coronary arteries in hypercholesterolemia and in atherosclerosis. *Circ. Res.* 64:900.

Siebenmann, R., Turina, M., Bühler, F.R. et al., 1990, Interaction between endothelin-1 and endothelium-derived relaxing factor in human arteries and veins. *Circ. Res.* 66 (4):1988.

Simon, B.C., Cunningham, L.D., Cohen R.A., 1990, Oxidized low density lipoproteins cause contraction and inhibit endothelium-dependent relaxation in the pig coronary artery. *J. Clin. Invest.* 86:75.

Standen, N.B., *et al.*, 1989, Hyperpolarizing vasodilators activate A sensitive K$^+$ channels in arterial smooth muscle. *Science* 245:177.

Stewart, D.J., Langleben, D., Cernacek, P., Cianflone, K., 1990, Endothelin release is inhibited by co-culture of endothelial cells with cells of vascular media. *Am. J. Physiol.* 259:H1928.

Suzuki, N., *et al.*, 1989, Immunoreactive endothelin-1 in plasma detected by a sandwich-type enzyme immunoassay. *J. Cardiovasc. Pharmacol.* 13 (Suppl.5):151.

Tanner, F.C., Tschudi, M.R., Lüscher, T.F., 1991, Endothelium, lipoproteins and atherosclerotic vascular disease. *Vasc. Med. Rev.* 2:161.

Tanner, F.C., Noll, G., Boulanger, C.M., Lüscher T.F., 1991, Oxidized native low density lipoproteins inhibit relaxations of porcine coronary arteries: Role of scavenger receptor and endothelium-derived nitric oxide. *Circulation* 83:2012.

Tare, M., *et al.*, 1990, Hyperpolarization and relaxation of arterial smooth muscle caused by NO derived from the endothelium. *Nature* 346:69.

Tesfamariam, B., Jakubowski, J.A., Cohen, R.A., 1989, Contraction of diabetic rabbit aorta due to endothelium-derived PGH$_2$/TXA$_2$. *Am. J. Physiol.* 257:1327.

Thom, S., Hughs, A., Sever, P.S., 1988, Endothelium dependent responses in human arteries In: Relaxing and Contracting Factors. Biological and Clinical Research. Edited by P.M. Vanhoutte. Clifton, NJ, *Humana Press* 511.

Tschudi, M.R., Criscione, L, Lüscher, T.F., 1991, Activation of the endothelial L-arginine pathway in rat coronary arteries: Effect of age and hypertension. *J. Hypertens.* (In Press).

Vita, J.A., *et al.*, 1990, Coronary vasomotor response to acetylcholine relates to risk factors for coronary artery disease. *Circulation* 81:491.

Warner, T.D., Mitchell, J.A., de Nucci, G., Vane, J.R., 1989, Endothelin-1 and endothelin-3 release EDRF from isolated perfused arterial vessels of the rat and rabbit. *J. Cardiovasc. Pharmacol.* 13 (Suppl 5):85.

Wood, K.S., Buga, G.M., Byrns, R.E., Ignarro, L.J., 1990, Vascular smooth muscle-derived relaxing factor (MDRF) and its close similarity to nitric oxide. *Biochem. Biophys. Res. Comm.* 170(1):80.

Yanagisawa, M., *et al.*, 1988, A novel potent vasoconstrictor peptide produced by vascular endothelial cells. *Nature* 332:411.

Yang, Z., Stulz, P., von Segesser, L., Bauer, E., Turina, M., Lüscher, T.F., 1991, Different interactions of platelets with arterial and venous coronary bypass vessels. *Lancet* 337:939.

Yang, Z., Siebenmann, R., Studer, M., Egloff, L., Lüscher, T.F., 1992, Similar endothelium-dependent relaxation, but enhanced contractility of the right gastroepiploic artery as compared with the internal mammary artery. *J. Thorac. Cardiovasc. Surg.* 104:459.

Yang, Z., *et al.*, 1991, Different activation of endothelial L-arginine and cyclooxygenase pathway in human internal mammary artery and saphenous vein. *Circ. Res.* 68:52.

Yang, Z., von Segesser, L., Stulz, P., Turina, M., Lüscher, T.F., 1992, Pulsatile stretch and platelet-derived growth factor (PDGF): Important mechanisms for coronary venous bypass graft disease. *Circulation* 86 (Suppl. I):I-84.

Yang, Z., Richard, V., von Seagesser, L., Bauer, E. , Stulz, P., Turina, M., Lüscher, T.F., 1990, Threshold concentrations of endothelin-1 potentiate contractions to norepinephrine and serotonin in human arteries: A new mechanism of vasospasm? *Circulation* 82:188.

Yang, Z., Bauer, E., von Segesser, L., Stulz, P., Turina, M., Lüscher, T.F., 1990, Different mobilization of calcium in endothelin-1-induced contractions in human arteries and veins: Effects of calcium antagonists. *J. Cardiovasc. Pharmacol.* 16:654.

Ylä-Herttuala, S., *et al.*, 1989, Evidence for the presence of oxidatively modified low-density lipoproteins in atherosclerotic lesions of rabbit and man. *J. Clin. Invest.* 84:1086.

Yokokawa, K., *et al.*, 1991, Endothelin-3 regulates endothelin-1 production in cultured human endothelial cells. *Hypertension* 18:304.

Yoshimoto, S., *et al.*, 1990, The role of cerebral microvessel endothelium in regulation of cerebral blood flow through production of endothelin-1 (Abstract). *J. Vasc. Med. Biol.* 2 (4) : 178.

ISOFORMS OF NITRIC OXIDE SYNTHASE AND THE NITRIC

OXIDE-CYCLIC GMP SIGNAL TRANSDUCTION SYSTEM

Ferid Murad, U. Förstermann, M. Nakane, J. Pollock,
H. Schmidt, T. Matsumoto, R. Tracey and W. Buechler

Abbott Laboratories
Abbott Park, IL
U.S.A.

Northwestern University
Chicago, IL
U.S.A.

From the work in our laboratory and subsequently other laboratories, it has been known for many years that cyclic GMP induces the relaxation of numerous smooth muscle preparations including vascular, airway and intestinal smooth muscle (*Katsuki and Murad, 1977; Katsuki et al., 1977; Murad et al., 1978; Murad et al., 1978; Rapaport and Murad, 1983; Murad, 1986*). Smooth muscle relaxation was the first physiological function clearly related to cyclic GMP synthesis. The proposed functions of cyclic GMP have expanded considerably since, as briefly discussed below.

The increase in cyclic GMP in smooth muscle preparations is associated with cyclic GMP-dependent protein kinase activation and altered phosphorylation of numerous endogenous smooth muscle proteins (*Rapaport et al., 1982; Fiscus et al., 1983; Rapoport et al., 1983*). The functions of most of these phosphoproteins have not been determined. Cyclic GMP can also decrease the activity of phospholipase C in vascular preparations and smooth muscle cell cultures as determined by decreased formation of inositol phosphates, and this effect also appears mediated by cyclic GMP-dependent protein kinase (*Rapoport, 1986; Hirata et al., 1990*). Both of these effects probably lower cytosolic free calcium concentrations resulting in the decreased activity of myosin light chain kinase, decreased phosphorylation of myosin light chain and relaxation (*Murad, 1986; Fiscus et al., 1983; Rapoport et al., 1983*). Cyclic GMP may alter calcium concentrations through other mechanisms as well and could also have effects that are mediated independently of cyclic GMP-dependent protein kinase activation. However, such pathways have not clearly been described in smooth muscle to date as they have in systems such as the retina (*Stryer, 1986*). Earlier reports (*Popescu et al., 1985*) that cyclic GMP can alter the activity of a membrane Ca^{++}-Mg^{++} ATPase through a cyclic GMP-dependent protein kinase mechanism have not been confirmed.

A considerable amount of data would suggest that cyclic GMP is a more potent relaxant of smooth muscle than is cyclic AMP. For example, the relaxant effects of cyclic nucleotide phosphodiesterase inhibitors are frequently associated with greater increases in cyclic GMP than cyclic AMP, and cyclic GMP analogues are generally more potent than cyclic AMP analogues with smooth muscle relaxation (*Katsuki and Murad, 1977; Rapoport and Murad, 1983*).

Other functions of cyclic GMP have included phototransduction in the retina

(*Stryer, 1986*), enterotoxin-induced intestinal secretion (*Hughes et al., 1978; Guerrant et al., 1980; Waldman and Murad, 1987*), inhibition of platelet aggregation (*Walter, 1989*) and a variety of less clearly defined effects (*for reviews see Waldman and Murad, 1987; Walter, 1989*).

An appreciation of the smooth muscle relaxant effects of cyclic GMP came from our earlier studies with azide, nitrite and numerous *nitrovasodilators* which increased cyclic GMP synthesis and relaxed a variety of smooth muscle preparations (*Katsuki and Murad, 1977; Katsuki et al., 1977; Kimura et al., 1975; Kimura et al., 1975*). These agents share a common feature in that they either liberate nitric oxide or can be enzymatically converted to this reactive free radical which activates soluble guanylyl cyclase (*Katsuki et al., 1977; Arnold et al., 1977*). These studies have since been confirmed by numerous laboratories working with a variety of tissue preparations (*see review Waldman and Murad, 1987*). We coined the term "*nitrovasodilators*" for this broad class of nitric oxide forming agents even though many of these agents do not possess a nitro or nitroso functionality and require oxidation and enzymatic conversion (*Murad et al., 1978; Murad et al., 1978; Rapoport and Murad, 1983; Murad, 1986; Rapoport et al., 1983; Waldman and Murad, 1987; Murad, 1990*). We believe this term has facilitated communication in this field.

It is rather uniformly believed that only the soluble isoform of guanylyl cyclase can be activated by nitrovasodilators. The soluble isoform is a heterodimer (*Kamisaki et al., 1986*) and each dimer appears to possess a catalytic domain from our cloning experiments (*Nakane et al., 1988; Nakane et al., 1990*). While the enzyme possesses a heme moiety which is thought to participate in the activation mechanism (*Gerzer at al., 1981; Ignarro et al., 1986; Lewicki et al, 1982*), the actual mechanism of activation has not been definitively determined since this requires large quantities of active protein for the appropriate physicochemical studies.

While soluble guanylyl cyclase is clearly activated by the nitric oxide generating nitrovasodilator agents, some of the particulate isoforms of guanylyl cyclase can also be activated. These include enzyme preparations from retina, intestinal epithelium and liver (*Waldman and Murad, 1987; Horio and Murad, 1991; Waldman et al., 1982*). The particulate isoforms of guanylyl cyclase are clearly different structurally from the soluble enzyme (*Waldman and Murad, 1987; Murad, 1989; Garbers, 1991*). However, some common regulatory features may be shared. Perhaps some of the particulate isoforms also contain a heme moiety, a transition metal or critical sulfhydryl groups that interact with NO to cause conformational changes in the enzyme and activation. Clearly, numerous experiments are yet to be conducted.

Another interesting feature of soluble guanylyl cyclase is that while each subunit appears to possess a catalytic domain, both subunits are required for catalytic activity and nitric oxide activation (*Nakane et al., 1990; Saheki et al., 1990*). Our site directed mutagenesis studies with soluble guanylyl cyclase are incomplete and inconclusive to date to clarify the complexities of the subunit interactions.

After finding that nitrovasodilators mediated their effects through nitric oxide production and increased cyclic GMP synthesis, we proposed that nitric oxide could be a natural endogenous agent whose formation from some endogenous precursor(s) could be hormonally regulated in order to explain hormonal regulation of cyclic GMP synthesis (*Murad et al., 1978; Murad et al., 1978*). Since nanamolar concentrations of NO are capable of activating guanylyl cyclase and increasing cyclic GMP, and all of the assays for NO and nitrite were several orders of magnitude less sensitive, it was not possible to actively test this hypothesis with definitive experiments. It took another decade and work from several independent directions before this hypothesis was proven.

Furchgott and his associates found that some vasorelaxants required the presence of the endothelium to cause the release of an endothelium derived relaxant factor and vascular relaxation (*Furchgott and Zawadski, 1980; Furchgott and Vanhoutte, 1989*). After hearing Furchgott present this work in 1980 at a seminar, we discussed the similarities of the actions of EDRF and nitrovasodilators, and I suggested that EDRF could mediate its effects via cyclic GMP. Unfortunately , a relocation of our laboratory in 1981 to California prevented such a collaboration from occurring. However, shortly thereafter we began to repeat Furchgott's experiments with endothelium-dependent vasodilators and found that the effects of EDRF were indeed mediated through increased cyclic GMP synthesis (*Rapoport and Murad, 1983; Murad,*

1986; Rapoport et al., 1982; Fiscus et al., 1983; Rapoport and Murad, 1983). Further-more, endothelium-dependent vasodilators increased cyclic GMP-dependent protein kinase activity and altered the phosphorylation of the same identical proteins identified on 2D gels as did nitrovasodilators including the decrease in myosin light chain phosphorylation (Rapoport et al., 1982; Fiscus et al., 1983; Rapoport et al., 1983; for review see Murad, 1986; Waldman and Murad, 1987). Thus, the pharmacological and biochemical effects of endothelium-dependent vasodilators and nitrovasodilators were virtually identical, and we came to view EDRF as an "endogenous nitrovasodilator" (Murad, 1986). Shortly thereafter, we again proposed that nitric oxide formation could be responsible for hormonally-induced cyclic GMP synthesis in some tissues (Murad et al., 1988; Murad, 1989). At about the same time, Furchgott (Furchgott, 1988) and Ignarro (Ignarro, 1987) proposed that EDRF could be nitric oxide. We and others have argued that EDRF, as originally defined by Furchgott, could not be NO per se but could be a precursor of NO or a NO complex or adduct that liberates NO to activate guanylyl cyclase (Murad et al., 1988; Murad, 1989). While this controversy has not been suitably resolved to date, more and more evidence suggest that EDRF is a complex that can liberate NO. While nitric oxide synthase can indeed convert arginine to citrulline and nitric oxide as discussed below, nitric oxide probably forms a complex with some endogenous material(s) before or after it is liberated from cells. Recent studies suggest that this nitric oxide complex may contain a thiol (Myers et al., 1990). Presumably, this complex or complexes behave as EDRF(s) and subsequently liberate(s) nitric oxide which activates guanylyl cyclase. This hypothesis is much more appealing to us, particularly if there were specific and/or selective transport and delivery mechanisms for this EDRF complex. The marked reactivity of the nitric oxide free radical with numerous substances (Braughler et al., 1979) seems to preclude NO itself being EDRF. Nitric oxide at the nanamolar concentrations that are required for this signal transduction process would probably fail to reach and activate soluble guanylyl cyclase in a neighboring targeted cell before colliding with other molecules to form an inactive complex. We demonstrated a number of years ago that the dose response curve for NO activation of purified guanylyl cyclase is markedly shifted to the right in the presence of albumin, thiols, sugars and other substances (Braughler et al., 1979). Unfortunately, the low concentrations of EDRF and its reactivity and short half life will require some novel challenging chemistry to identify the specific structure of EDRF which undoubtedly contains NO. It is not anticipated that the precise structure of EDRF will be definitively proven in the near future.

Quite independently, Hibbs et al., found that macrophages could produce nitric oxide and nitrite which was associated with their cytotoxic properties (Hibbs et al., 1990). This laboratory also reported that arginine analogues could prevent NO/nitrite formation. This observation was rapidly recognized by many laboratories that proceeded to show that arginine was converted to citrulline and nitric oxide in numerous cell-free and intact cell systems including macrophages, leukocytes, endothelial cells, brain, etc. This enzyme has been referred to as NO synthase, EDRF synthase, or guanylyl cyclase activating factor (GAF) synthase (Bredt and Snyder, 1990; Schmidt et al., 1991; Stuehr et al., 1991; Garthwaite et al., 1988).

Today we know that there are at least four isoforms of this enzyme which are found in numerous tissues (Table 1). (Forstermann et al., 1991; Forstermann et al., 1991; Forstermann et al., 1990). The enzyme may be soluble of particulate and it may be constitutive or inducible with endotoxin and cytokines. The constitutive isoforms are generally regulated by Ca^{++}-calmodulin while the inducible forms are not. Most of the isoforms require NADPH, tetrahydrobiopterin, FAD and FMN (see Table 1). Several isoforms have been purified to homogeneity (Bredt and Snyder, 1990; Schmidt et al., 1991; Stuehr et al., 1991; Yui et al., 1991; Pollock et al., 1991) and antibodies have been generated (Schmidt et al., 1992a; Schmidt et al., submitted; Bredt et al., 1990; Pollock et al., 1992). The soluble brain enzyme (Type 1) has been cloned (Bredt et al., 1991) and cloning activity continues with the other isoforms. In addition to the regulation of catalytic activity by the cofactors listed above and Ca^{++}/calmodulin, there is some evidence that phosphorylation of the enzyme by various kinases can increase and/or decrease activity (Nakane et al., 1991). The regulatory significance, if any, with phosphorylation of the enzyme to control catalytic activity is not presently known. Experiments to date have been conducted with cell-free systems that are obviously more difficult to correlate with more physiological intact cell or tissue systems.

TABLE 1: Isoforms of NO synthase.

Type	Cosubstrates Co-factors	Regulated by	M_r	Present in
I a (soluble) FAC/FMN	NADPH, BH$_4$,	Ca^{2+}/calmodulin	155 kDa	brain,cerebellum N1E-115 neuroblastoma cells
1 b (soluble)	NADPH	Ca^{2+}/calmodulin	135 kDa	endothelial cells
I c (soluble)	NADPH, BH$_4$ FAD	Ca^{2+} (not calmodulin)	150 KdA	neutrophils
II (soluble)	NADPH BH$_4$ FAD/FMN	unknown (induced by endotoxin/ cytokines)	125 kDa	macrophages and many other induced tissues
III (particulate)	NADPH BH$_4$ FAD/FMN	Ca^{2+}/calmodulin	135 kDa	endothelial cells
IV (particulate)	NADPH	unknown (induced by endotoxin/ cytokines)	?	macrophages and many other induced tissues

All isoenzymes use L-arginine as a substrate and all are inhibited by N^G-methyl-L-arginine and N^G-nitro-L-arginine.

Immunohistochemical localization of the Type I and Type III isoforms with selective polyclonal and monoclonal antibodies that we and others have generated demonstrate a rather ubiquitous distribution of these isoforms in numerous tissues (*Schmidt et al., 1992a; Bredt et al., 1990; Pollock et al., 1992*).

Our polyclonal antibodies to the brain Type I enzyme are quite selective and do not recognize the Type III enzyme from endothelial cells. Monoclonal antibodies to endothelial Type III enzyme may be selective for the Type III enzyme or may cross-react with common epitopes shared by the Type I and Type III enzyme (*Pollock et al., 1992*).

We, and others, have also found that nonadrenergic-noncholinergic (NANC) innervated tissues, such as bovine retractor penis and rat anococcygeus, possess a brain Type I NO synthase activity which we have partially purified and characterized (*Mitchell et al., 1991; Sheng et al., 1992*). Furthermore, nerve fibers in these tissues, as well as gastrointestinal tissues and gastrointestinal smooth muscle, demonstrate immunohistochemical staining with antibody to the Type I brain NO synthase (*Sheng et al., submitted*). Thus, nitric oxide is probably one of the neurotransmitters of NANC neurons and the presence of NANC or "nitrinergic neurons" is quite ubiquitous in the peripheral as well as the central nervous system.

From our work and the work of others, it is now quite apparent that the nitric oxide-cyclic GMP system can function as an intracellular or an intercellular signal transduction system (*Murad, 1989; Ishii et al., 1989; Ishii et al., 1989*). If a specific cell possesses both nitric oxide synthase and an isoform of guanylyl cyclase that is

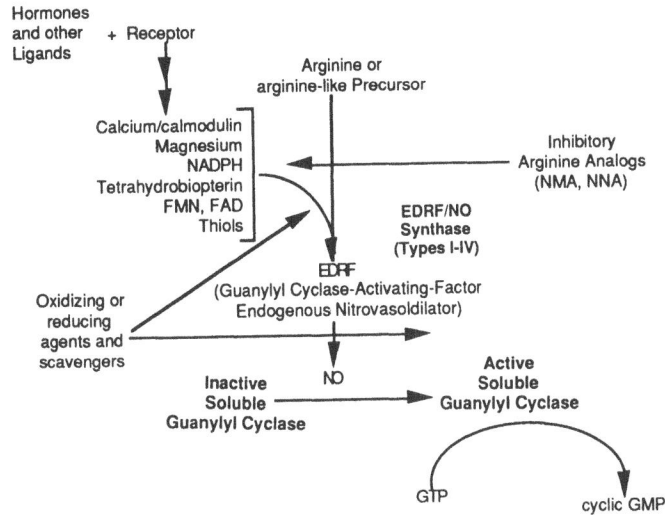

FIGURE 1: The nitric oxide\cyclic GMP signal transduction pathway. The nitric oxide formed from the oxidation of the quanidino nitrogen of arginine can act as an intracellular or intercellular messenger to regulate cyclic GMP synthesis.

activatable with NO, then cyclic GMP levels in that cell can be regulated through the process (Figure 1). If, however, the cell which generates NO lacks an activatable isoform of guanylyl cyclase, NO or a complex of NO, which is liberated from the producing or donor cell, can activate guanylyl cyclase in a neighboring cell to increase cyclic GMP synthesis. In the latter scenario, NO or its carrier complex behaves as a paracrine substance or autacoid. Interestingly, the liberated extracellular NO can also loop back and increase cyclic GMP synthesis in the cell of origin. This is best demonstrated by the inhibitory effects of hemoglobin on agonist-induced cyclic GMP accumulation in homogenous cell culture systems where the hormone or agonist effects on cyclic GMP are mediated by NO. Presumably, hemoglobin would not be permeable and could only trap or scavenge extracellular NO to account for its ability to decrease hormonally-induced cyclic GMP increases in homogenous cell populations. There is no evidence that NO can act as an endocrine substance to increase cyclic GMP synthesis in a distant target cell population. However, complexes of NO that would be relatively stable and later liberate NO at a distant site, could most certainly be viewed as endocrinologic agents (hormones or autacoids). We suspect that appropriately designed experiments in the future will also support this role for NO as an endocrinologic agent that can also function at a distance similar to classical hormones.

Indeed, we believe that nitric oxide should be added to the list of agents that can function as a neurotransmitter, paracrine substance and autacoid or hormone. It can also be viewed as an intracellular, as well as intercellular, messenger. To date, no substance or class of substances have played such a diverse role in intracellular and intercellular signal transduction. Thus, NO appears to be a unique and simple molecule with diverse functions in signal transduction.

ACKNOWLEDGEMENTS

Numerous trainees and collaborators have participated in our studies with the nitric oxide-cyclic GMP system in a wide variety of tissues and systems during the past 18 years. The discussion, debates and progress have been rewarding to all of us. Many of the studies summarized here were supported with grants from the NIH and many other agencies.

REFERENCES

Arnold, W.P., Mittal, C.K., Katsuki, S. and Murad, F., 1977, Nitric oxide activates guanylate cyclase and increases guanosine 3´, 5´-monophosphate levels in various tissue preparations. *Proc. Nat. Acad. Sci. USA* 74:3203-3207.

Braughler, J.M., Mittal, C.K. and Murad, F., 1979, Effects of thiols, sugars and proteins on nitric oxide activation of guanylate cyclase. *J. Biol. Chem.* 254:12450-12454.

Bredt, D.S. and Snyder, S.H., 1990, Isolation of nitric oxide synthetase, a calmodulin requiring enzyme. *Proc. Nat. Acad. Sci. USA* 85:682-685.

Bredt, D.S., Hwang, P.M. and Snyder, S.H., 1990, Localization of nitric oxide synthase indicating a neu ral role for nitric oxide. *Nature* 347:768-770.

Bredt, D.S., Hwang, P.M., Glatt, C.G., Lawenstein, C., Reed, R.R. and Synder, S.H., 1991, Cloned and expressed nitric oxide synthase structurally resembles cytochrome P-450 reductase. *Nature* 351:714-718.

Fiscus, R.R., Rapoport, R.M. and Murad, F., 1983, Endothelium-dependent and nitrovasodilator-induced activation of cyclic GMP-dependent protein kinase in rat aorta. *J. Cyclic Nucl. and Protein Phosphor. Res.* 9:415-425.

Förstermann, U., Gorsky, L., Pollock, J.S., Schmidt, H.H.H.W., Ishii, K., Heller, M. and Murad, F., 19 90, Subcellular localization and regulation of the enzymes responsible for EDRF synthesis in endothelial cells and N1E 115 neuroblastoma cells. *Eur. J. Pharmacol.* 183:1625-1626.

Förstermann, U., Pollock, J., Schmidt. H.H.H.W., Heller, M. and Murad, F., 1991, Calmodulin-depen dent endothelium-derived relaxing factor/nitric oxide synthase activity is present in the partic-ulate and cytosolic fractions of bovine aortic endothelial cells. *Proc. Nat. Acad. Sci.* 88:1788-1792.

Förstermann, U., Schmidt, H.H.H.W., Pollock, J.S., Sheng, H., Mitchell, J.A., Warner, T.D., Nakane, M. and Murad, F., 1991, Isoforms of EDRF/NO synthase: Characterization and purification from different cell types. *Biochem. Pharmacol.* 41:1849-1857.

Furchgott, R.F. and Vanhoutte, P.M., 1989, Endothelium-derived relaxing and contracting factors. *FASEB J.* 3:2007-2018.

Furchgott, R.F. and Zawadski, J.V., 1980, The obligatory role of endothelial cells in the relaxation of arterial smooth muscle to acetylcholine. *Nature* 288:373-376.

Furchgott, R.F., 1988, Studies on relaxation of rabbit aorta by sodium nitrite: The basis for the propo-sal that acid-activatable inhibitory factor from bovine retractor penis is organic nitrite and EDRF is nitric oxide. In: Vanhoutte, P.M. (ed); Vasodilation: Vascular Smooth Muscle, Peptides, Autonomic Nerves and Endothelium. New York Raven, 401-414.

Garbers, D.L., 1991, The guanylyl cyclase-receptor family. *Can. J. Physiol. Pharmacol.* 69:1618-1621.

Garthwaite, J., Charles, S.L. and Chess-Williams, R., 1988, Endothelium-derived relaxing factor release on activation of NMDA receptors suggests role as intercellular messenger in the brain. *Nature* 336:385-387.

Gerzer, R., Bohme, E., Hoffman, F. and Schultz, G., 1981, Soluble guanylate cyclase purified from bovine lung contains heme and copper. *FEBS Lett.* 132:71-74.

Guerrant, R.L., Hughes, J.M., Chang, B., Robertson, D.C. and Murad, F., 1980, Activation of intesti-nal guanylate cyclase by heat stable enterotoxin of Escherichia Coli: Studies of tissue specificity, potential receptors and intermediates. *J. Infec. Dis.* 142:220-228.

Hibbs, J.R., Taintor, R.R. and Varrin, Z., 1987, Macrophage cytotoxicity: Role for L-arginine deimi-nase and imino nitrogen oxidation to nitrite. *Science* 235:473-476.

Hirata, M., Kohse, K., Chang, C.H., Ikebe, T. and Murad, F., 1990, Mechanism of cyclic GMP inhibi-tion of inositol phosphate formation in rat aorta segments and cultured bovine aortic smooth muscle cells. *J. Biol. Chem.* 265:1268-1273.

Horio, Y. and Murad, F., 1991, Purification of guanylyl cyclase from bovine rod outer segments. *Bioch im. Biophys. Acta*, 1133:81-88.

Hughes, J., Murad, F., Chang, B. and Guerrant, R., 1978, The role of cyclic GMP in the mechanism of action of the heat-stable enterotoxin of E. Coli. *Nature* 271:755-756.

Ignarro, L.J., Adams, J.B., Horwitz, P.M. and Wood, K.S., 1986, Activation of soluble guanylate cy-clase by NO-hemeproteins involves NP-heme exchange: comparison of heme containing and heme deficient enzymes. *J. Biol. Chem.* 261:4997-5002.

Ignarro, L.J., Buga, G.M., Wood, K.S., Byrns, R.E. and Chaudhuri, G., 1987, Endothelium-derived re-laxing factor produced and released from artery and vein is nitric oxide. *Proc. Nat. Acad. Sci.* 84:9265-9269.

Ishii, K., Chang, B., Kerwin, J.F., Wagenaar, F.L., Huang, Z.J. and Murad, F., 1991, Formation of EDRF in porcine kidney epithelial LLC-PK₁ cells: An intra- and intercellular messenger for activation of soluble guanylate cyclase. *J. PET.* 256:38-53.

Ishii, K., Gorsky, L., Förstermann, U. and Murad, F., 1989, Endothelium-derived relaxing factor (EDRF): The endogenous activator of soluble guanylate cyclase in various types of cells. *J. Applied Cardiology* 4:505-512.

Kamisaki, Y., Saheki, S., Nakane, M., Palmieri, J., Kuno, T., Chang, B., Waldman, S.A. and Murad, F., 1986, Soluble guanylate cyclase from rat lung exists as a heterodimer. *J. Biol. Chem.* 261:7236-7241.

Katsuki, S. and Murad, F., 1977, Regulation of cyclic 3´, 5´-adenosine monophosphate and cyclic 3´, 5´-guanosine monophosphate levels and contractility in bovine tracheal smooth muscle. *Molecular Pharmacology* 13:330-341.

Katsuki, S., Arnold, W.P. and Murad, F., 1977, Effect of sodium nitroprusside, nitroglycerin and sodium azi-de on levels of cyclic nucleotides and mechanical activity of various tissues. *J. Cyclic Nucl. Res.* 3:239-247.

Katsuki, S., Arnold, W., Mittal, C.K. and Murad, F., 1977, Stimulation of guanylate cyclase by sodium nitroprusside, nitroglycerin and nitric oxide in various tissue preparations and comparison to the effects of sodium azide and hydroxylamine. *J. Cyclic Nucl. Res.* 3:23-35.

Kimura, H., Mittal, C.K. and Murad, F., 1975, Activation of guanylate cyclase from rat liver and other tissues with sodium azide. *J. Biol. Chem.* 250:8016-8022.

Kimura, H., Mittal, C.K. and Murad, F., 1975, Increases in cyclic GMP levels in brain and liver with sodium azide, an activator of guanylate cyclase. *Nature* 257:700-702.

Lewicki, J.A., Grandwein, H.J., Mittal, C.K., Arnold, W.P. and Murad, F., 1982, Properties of purified soluble guanylate cyclase activated by nitric oxide and sodium nitroprusside. *J. Cyclic Nucl. Res.* 8:17-25.

Mitchell, J.A., Sheng, H., Förstermann, U. and Murad, F., 1991, Characterization of nitric oxide synthases in non-adrenergic, non-cholinergic nerve containing tissue from the rat anococcygenus muscle. *Brit. J. Pharmacol.* 104:289-291.

Mittal, C.K., Kimura, H. and Murad, F., 1975, Requirement for a macromolecular factor for sodium azide activation of guanylate cyclase. *J. Cyclic Nucl. Res.* 1:261-269.

Murad, F. and Ishii, K., 1990, Hormonal regulation of the different isoforms of guanylate cyclase: EDRF is a ubiquitous activator of soluble guanylate cyclase. In: Rubanyi, G.M. and Vanhoutte, P. (eds); Proc. of the First Internat. Symp. on Endothelium-Derived Vasoactive Factors. Philadelphia, PA, Karger, Basel, 151-165.

Murad, F., 1986, Cyclic guanosine monophosphate as a mediator of vasodilation. *J. Clin. Invest.* 78:1-5.

Murad, F., 1990, Drugs used in the treatment of angina: organic nitrites, calcium channel blockers and β-adrenergic antagonists. In Gilman, A.G., Rall, T.W., Nies, A. and Taylor, P. (eds); *Pharmacological Basis of Therapeutics, VII Edition,* 32:764-783.

Murad, F., 1989, Mechanisms for hormonal regulation of the different isoforms of guanylate cylase. In: Gehring, Y., Helmreich, E. and Schults, R. (eds); *Molecular Mechanisms of Hormone Action.* Springer Heidelberg, 186-194.

Murad, F., 1989, Modulation of the guanylate cyclase-cGMP system by vasodilators and the role of free radicals as second messengers. In: Catravas, J.D., Gillis, C.N. and Ryan, U.S. (eds); *Vascular Endothelium,* Plenum Pub., 157-164.

Murad, F., Ishii, K., Förstermann, U., Gorsky, L., Kerwin, J., Pollock, J. and Heller, M., 1990, EDRF is an intracellular second messenger and autacoid to regulate cyclic GMP synthesis in many cells. *Adv. Cyclic. Nucl. Res.* 24:441-448.

Murad, F., Leitman, D., Waldman, S., Chang, C.H., Hirata, J., Kohse, K., 1988, Effects of nitrovasodilators, endothelium-dependent vasodilators and atrial peptides on cGMP. *Proc. Cold Spring Harbor Symposium on Quantitative Biology, Signal Transduction* 53:1005-1009.

Murad, F., Mittal, C.K., Arnold, W.P. and Braughler, J.M., 1978, Effects of nitro-compound smooth muscle relaxants and other materials on cyclic GMP metabolism. In: J.C. Stocklet (ed); *Advances in Pharmacology and Therapeutics, Vol 3 Ions, Cyclic Nucleotides, Cholinergy,* Pergamon Press, New York, 123-132.

Murad, F., Mittal, C.K., Arnold, W.P., Katsuki, S. and Kimura, H., 1978, Guanylate cyclase: Activation by azide, nitro compounds, nitric oxide, and hydroxyl radical and inhibition by hemoglobin and myoglobin. *Adv. Cyclic Nucl. Res.* 9:145-158.

Myers, P.R., Minor, R.L., Guerro, R., Bates, J.N. and Harrison, D.G., 1990, Vasorelaxant properties of the endothelium-derived relaxing factor more closely resemble S-nitrosocysteine than nitric oxide. *Nature* 345:161-163.

Nakane, M., Arai, K., Saheki, S., Kuno, T., Buechler, W. and Murad, F., 1990, Molecular cloning and expression of cDNAs coding for soluble guanylate cyclase from rat lung. *J. Biol. Chem.* 265:1-6841-16845.

Nakane, M., Mitchell, J.A., Förstermann, U. and Murad, F., 1991, Phosphorylation by calcium calmodulin-dependent protein kinase II and protein kinase C modulates the activity of nitric oxide synthase. *Biochem. Biophys. Res. Com.* 180:1396-1402.

Nakane, M., Saheki, S., Kuno, T., Ishii, K., Deguchi, T. and Murad, F., 1988, Molecular cloning of a cDNA coding for 70 kilodalton subunit of soluble guanylate cyclase from rat lung. *Biochem. Biophys. Res. Commun.* 157:1139-1147.

Pollock, J.S., Förstermann, U., Mitchell, J.A., Warner, T.D., Schmidt, H.H.H.W., Nakane, M. and Murad, F., 1991, Purification and characterization of EDRF particulate synthase from cultured and native bovine aortic endothelial cells. *Proc. Nat. Acad. Sci. USA*, 88:10480-10484.

Pollock, J.S., Nakane, M., Förstermann, U. and Murad, F., 1992, Characterization of monoclonal anti bodies to Type III nitric oxide (NO) synthase. Presented as ASBMB/Biophysical Society, Houston, Texas, February.

Popescu, L.M., Panoiu, C., Hinescu, M. and Nutu, O., 1985, The mechanism of cGMP-induced relaxation in
vascular smooth muscle. *Eur. J. Pharmacol.* 107:393-394.

Rapoport, R.M. and Murad, F., 1983, Agonist-induced endothelial-dependent relaxation in rat thoracic aorta may be mediated through cyclic GMP. *Circ. Res.* 52:352-357.

Rapoport, R.M. and Murad, F., 1983, Endothelium-dependent and nitrovasodilator-induced relaxation of vascular smooth muscle: Role for cyclic GMP. *J. Cyclic Nucl. and Protein Phosphor. Res.* 9:281-296.

Rapoport, R.M., 1986, Cyclic guanosine monophosphate inhibition of contraction may be mediated through inhibition of phosphotidylinositol hydrolysis in rat aorta. *Circ. Res.* 58:407-410.

Rapoport, R.M., Draznin, M. and Murad, F., 1982, Sodium nitroprusside-induced protein phosphorylation in intact rat aorta is mimicked by 8-bromo-cyclic GMP. *Proc. Nat. Acad. Sci. USA* 79:6470-6474.

Rapoport, R.M., Draznin, M.D. and Murad, F., 1983, Endothelium-dependent vasodilator-and-nitro vasodilator-induced relaxation may be mediated through cyclic GMP formation and cyclic GMP-dependent protein phosphorylation. *Trans. Assoc. Amer. Phys.* 96:19-30.

Saheki, S., Kuno, T., Takeuchi, N. and Murad, F., 1990, Radiation inactivation target size analysis of soluble guanylate cyclase. *Biochim. Biophys. Acta.* 1051:306-309.

Schmidt, H.H.H.W., Gagne, G.D., Nakane, M., Pollock, J.S., Förstermann, U., Miller, M.F. and Murad, F., 1992a, Mapping of neural NO synthase in the rat suggests frequent colocation with NADPH Diaphorase but not soluble guanylylcyclase and novel paraneural functions for nitrinergic signal transduction. *J. Histochem. Cytochem.* 40:1439-1456.

Schmidt, H.H.H.W., Pollock, J., Nakane, M., Gorsky, L., Förstermann, U., Heller, M. and Murad, F., 1991, Purification of a soluble isoform of guanylyl cyclase-activating factor synthase. *Proc. Nat. Acad. Sci.* 88:365-369.

Schmidt, H.H.H.W., Warner, T., Ishii, K., Sheng, H. and Murad, F., 1992b, Insulin secretion in pancreatic β cells caused by L-arginine derived nitric oxide. *Science* 255:721-723.

Sheng, H., Nakane, M., Schmidt, H.H.H.W., Mitchell, J., Pollock, J., Förstermann, U. and Murad, F., 1992, Characterization and localization of nitric oxide synthase in non-adrenergic non-cholinergic nerves from bovine retractor penis muscles. *Brit. J. Pharmacol.* 106:768-773.

Stryer, L., 1986, Cyclic GMP cascade of vision. *Annu. Rev. Neuroscience* 9:87-119.

Stuehr, D.J., Cho, H.J., Kwon, N.S., Wise, M.F. and Nathan, C.F., 1991, Purification and characterization of the cytokine-induced macrophage nitric oxide synthase: An FAC-and-FMN containing Flavoprotein. *Proc. Nat. Acad. Sci. USA* 88:7773-7777.

Waldman, S.A. and Murad, F., 1987, Cyclic GMP synthesis and function. *Pharm. Rev.* 39:163-196.

Waldman, S.A., Lewicki, J.A., Brandwein, H.J. and Murad, F., 1982, Partial purification and characterization of particulate guanylate cyclase from rat liver after solubilization with trypsin. *J. Cyclic Nuc. Res.* 8:359-370.

Walter, U., 1989, Physiological role of cGMP-dependent protein kinase in the cardiovas-cular system. *Rev. Physiol. Biochem. Pharmacol.* 113:41-48.

Yui, Y., Hattori, R., Kosuga, K., Eizaiva, H., Hiki, K., Ohkawa, S., Ohnishi, K., Terao, S. and Kawai, C., 1991, Calmodulin-independent nitric oxide synthase from rat polymorphonucleo neutrophils. *J. Biol. Chem.* 266:3369-3371.

REGULATION OF ANGIOGENESIS VIA PROTEIN KINASE C

Michael E. Maragoudakis, Nikos E. Tsopanoglou
and George Haralabopoulos

Department of Pharmacology
University of Patras Medical School
261 10 Patras
Greece

INTRODUCTION

Phorbol esters induce many effects in the cell via activation of protein kinase C (PKC) (*Castange et al., 1982*) and promote tumor formation (*Niedel et al., 1983; Blumberg, 1980*). The exact mechanism of tumor promotion is not understood. It has been suggested that stimulation of angiogenesis might be responsible for this effect (*Rifkin et al., 1981*). Indeed, studies in the chick chorioallantoic membrane (CAM) system have shown that 4-β-Phorbol-12-myristate-13-acetate (PMA) induced angiogenesis (*Morris et al., 1988; Maragoudakis, 1992*) and angiogenesis is considered an essential step for growth and metastasis of solid tumors (*Folkman, 1985*). This angiogenic effect of PMA was attributed by Morris *et al.* (1988) to the release of angiogenic factors rather than direct stimulation of endothelial cell proliferation.

In this report, we investigated the role of PMA on angiogenesis in relation to PKC. The involvement of PKC in the regulation of angiogenesis was established using the natural stimulator of PKC, diacylglycerol (DG) (*Boni and Rando, 1985*) and two protein kinase C inhibitors: H7 [1-(5-isoquinoline-sulfonyl)-2-methyl piperazine)] (*Hidaka et al., 1984*) and RO-318220 (*Davis et al., 1989; Linden and Conor, 1991*).

A newly described anti-angiogenic and antitumor agent D609 (tricyclodecan-9-yl-xanthogenate) (*Maragoudakis et al., 1990*) was shown to act like a PKC inhibitor in the CAM system.

In addition, the stimulatory effect of thrombin on angiogenesis reported previously (*Tsopanoglou et al., 1993*) is reversed by all the aforementioned inhibitors of PKC, indicating a pivotal role of PKC activation in the regulation of angiogenesis.

MATERIALS AND METHODS

PMA, H7, 1, 2-dioctanoyl-sn-glycerol (DG), collagenase type IV, cortisone acetate and other reagents were obtained from SIGMA Chemical Company (*London, England*), L-[U-^{14}C]-proline (270 μC/μmole) was obtained from New England Nuclear (Boston, MA). Plastic discs used were 13 mm round tissue culture coverslips from Nunc Inc. (Neperville, II). RO 318220 was a gift from Roche Products Ltd. D609 was a gift from Dr. Schatton, Merz and Co. (Frankfurt, Germany). Human a-thrombin was a gift from Dr. J. W. Fenton II. Fresh fertilized eggs were obtained from Ioannina, Greece and kept at 10^0C before incubation at 37^0C.

Biological assay for angiogenesis: The CAM method was used with certain modifications (*Folkman, 1985*). The test materials and 0.5 μCi [U-^{14}C] labelled proline

Table I: Effect of promoters and inhibitors of PKC on angio-genesis in the CAM

Promoter or Inhibitor	Collagenous Protein Biosynthesis	Vascular Density
	% Change from Control	
PMA (60ng/disc)	+50±9 (n=18)	+34±6 (n=5)
DG (10μg/disc)	+34±5 (n=20)	+34±5 (n=5)
H7 (30 μg/disc)	-38±3 (n=27)	-28±3 (n=10)
R0318220 (10 μg/disc)	-47±5 (n=10)	---
D609 (25 μg/disc)	-55±3 (n=38)	-38±4 (n=14)

Results are expressed as mean ± SE. The control values for collagenous protein synthesis was 15 ± 1x10±-3± cpm/mg protein, n=113 and vascular density was 131 ± 5 vessels, n=34.

were placed on the sterile discs and were allowed to dry under sterile conditions. Control discs containing equal amounts of radiolabelled proline and an appropriate volume of the vehicle, where the test material was dissolved, were also placed on the CAM about one cm away from the disc containing the test material. A sterile solution of cortisone acetate corresponding to 100 μg/disc was added to all the discs to prevent an inflammatory response. The loaded and dried discs were inverted and placed on the surface of the CAM on day 9. The windows were covered again with sterile cellophane tape and the eggs were returned to the incubator until day 11. After that the determination of collagenous protein biosynthesis was performed as described previously (*Maragoudakis et al., 1988*).

Morphological evaluation of angiogenesis: Eggs were treated as above with the exception that radiolabelled proline was not added. At day 11 the eggs were flooded with 10% buffered formalin at 37°C. The plastic discs were then removed and the eggs

Table II: Reversal of the effect of PMA and DG by PKC inhibitors

Combinations	Collagenous Protein Biosynthesis	Vascular Density
	% Change from Control	
PMA (60ng)+ H7 (30μG)	-46±6 (n=6)	-43±6 (n=5)
PMA (60ng)+ R0318220 (30 μg)	-43±6 (n=8)	---
PMA (60ng)+ D609 (25 μg)	-54±4 (n=12)	-43±6 (n=5)
DG (10ng) + H7 (30μg)	-32±7 (n=5)	-28 ± 6 (n=5)
DG (10ng) + R0318220 (10μg)	-47 ± 7 (n=8)	---
DG (10ng) + DG609 (25μg)	-61 ± 4 (n=6)	-42±5 (n=5)

Results are expressed as mean ± SE. The control values for collagenous protein synthesis was 12 ± 1x10³ cpm/mg protein, n=45 and vascular density was 135 ± 6 vessels, n=25

Table III: Reversal of a-thrombin stimulated angiogenesis by PKC inhibitors

Combinations	Collagenous Protein Biosynthesis	Vascular Density
	% Change from Control	
a-thrombin (1 iu/disc)	+77 ± 8 (n=59)	+41 ± 6 (n=6)
a-thr (1 iu) + H7 (30μg)	-38 ± 4 (n=12)	---
a-thr (1 iu) + R0318220 (10μg)	-15 ± 14 (n=6)	---
a-thr (1 iu) + D609 (25μg)	-48±6 (n=12)	---

Results are expressed as mean ± SE. The control values for collagenous protein synthesis was $15 \pm 1 \times 10^{-3}$, n = 89 and vascular density was 139 ± 7 vessels, n=6.

were kept at 37^0C until dissection. A large area around the disc was then removed, placed on a glass slide and stretched gently to its natural size. The vascular density index was measured by the method of Harris-Hooker et al. (1983), by counting the number of vessels intersecting each concentric circle of 4, 5 and 6 mm in diameter separately and then adding the vessels intersecting all three circles.

CALCULATIONS AND STATISTICS

For each egg, total collagenous protein biosynthesis and vascular density under the disc containing the test material were expressed as % of that under the control disc. The results were analyzed by paired t-test, and *n* signifies the number of eggs used for each treatment.

RESULTS AND DISCUSSION

Effect of promoters and inhibitors of PKC on collagenous protein synthesis and angiogenesis in the CAM system: It was shown previously that collagenous protein biosynthesis is a reliable biochemical index of angiogenesis (*Maragoudakis et al., 1988, 1992*). This method was used to evaluate the role of PKC in the regulation of angiogenesis. As shown in Table I the presence of PMA and DG caused stimulation in the rate of collagenous protein biosynthesis with a comeasurable increase in vascular density of the CAM. On the contrary, the inhibitors of PKC, H7 or RO 318220 caused an inhibition as compared to controls. The same effect was noted with the antitumor agent D609. All the aforementioned effects of these agents are dose-dependent (Data not shown).

When the activators (DG or PMA) were combined with the PKC inhibitors H7 and RO 318220 as shown in Table II, the rates of collagenous protein synthesis and angiogenesis were decreased to levels below the controls (without any additions). Similar results were obtained with D609 (Table II).

These results indicate that both DG, which is the natural activator of PKC, and the tumor promoter PMA, promote angiogenesis via a mechanism involving activation of PKC. Inhibitors of PKC such as H7 or RO 318220 not only prevent the stimulation of angiogenesis induced by PMA or DG but also reduce the rate of angiogenesis to levels below the control (basal angiogenesis). It appears, therefore, that PKC activity modulates the level of angiogenesis both under normal physiological conditions and also under conditions where angiogenesis is stimulated.

It is of interest in this respect that inhibitors of PKC reverse also the stimulatory effect of a-thrombin on angiogenesis (*Tsopanoglou et al.,* 1993) (Table III).

Both inhibitors of PKC, RO 318220 and H7, as well as D609, abolish the stimulatory effect of a-thrombin and depress angiogenesis below the control levels of basal angiogenesis.

This indicates that angiogenesis promoted by a variety of agents such as PMA, DG, a-thrombin as well as the basal (unstimulated) angiogenesis in the CAM system is regulated to a large extent via PKC. D609, which has been shown previously by Müller-Decker et al., 1988 to prevent phosphorylation of certain target proteins by PKC, exerts its antiangiogenic and antitumor effect (Maragoudakis et al., 1990) via inhibition of PKC. Comparison of D609 with R0318220 with respect to their inhibitory effect on angiogenesis indicate that D609 may be a potent inhibitor of PKC and, as such, a novel tool in elucidating the role of PKC in other systems.

In conclusion, angiogenesis is blocked to a large extent by inhibitors of PKC and stimulated by activators of PKC pointing to a pivotal role of this enzyme in the regulation of this important physiological process. It is likely, therefore, that activators or inhibitors of PKC may prove useful in modulating angiogenesis in disease states where this is the desirable therapeutic goal.

ACKNOWLEDGEMENTS

This work was supported by a grant from the Ministry of Industry, Energy and Technology of Greece. We thank Ms. Anna Marmara for typing of the manuscript.

REFERENCES

Blumberg, P.M., 1980, *In vitro* studies on the mode of action of the phorbol esters, potent tumor promoters. *CRC Crit. Rev. Toxicol.* 8:199-234.

Boni, L.T., and Rando, R.R., 1985, The nature of protein kinase C activation by physically defined phospho-lipid vesicles and diacylglycerols. *J. Biol. Chem.* 260:1081910825.

Castagna, M., Takai, Y., Kaibuchi, K., Sano, K., Kikkawa, U., and Nishizuka, Y., 1982, Direct activation of calcium-activated phospholipid - dependent protein kinase by tumor-promoting phorbol esters. *J. Biol. Chem.* 257:7845-7851.

Davis, P.D., Hill, C.H., Keech, E., Lawton, G., Nixon, J.S., Sedgwick, A.D., Wadsworth, J., Westma cott, D., and Wilkinson, S.E., 1989, Potent selective inhibitors of protein kinase C. *FEBS Lett.* 259(1): 61-63.

Folkman, J., 1983, Tumor angiogenesis. *Adv. Cancer Res.* 43:172-203.

Harris-Hooker, S. A., Gajdusek, S. M.,Wight, T. N. and Schwartz, S. M., 1983, Neo vascularization responses induced by cultured aortic endothelial cells. *J. Cell Physiol.* 114:302-310.

Hidaka, H., Inagaki, M., Kawamoto, S.and Susaki, Y., 1984, Isoquinolinesulfonamides, novel and potent inhibitors of cyclic nucleotide dependent protein kinase and protein kinase C. *Biochemistry* 23:5036-5041.

Linden, D.J., and Connor, J.A., 1991, Participation of postsynaptic PKC in cerebellar long-term depression in culture. *Science* 254:1656-1659.

Maragoudakis, M.E., 1992, Basement membrane biosynthesis as a target for developing angiogenesis inhibitors with antitumor properties. In: Cell and Molecular Biology of Basement Membranes in Health and Disease. N. Kefalides, ed., *Kidney International* (in press).

Maragoudakis, M.E., Panoutsacopoulou, M. and Sarmonika, M., 1988, Rate of basement membrane biosynthesis as an index to angiogenesis. *Tissue and Cell* 20(4):531-539.

Maragoudakis, M.E., Missirlis, E., Sarmonika, M., Panoutsacopoulou, M. and Karakiulakis, G., 1990, Basement membrane biosynthesis as a target to tumor therapy. *J. Pharmacol. Exp. Ther.* 252:753-757.

Morris, P.B., Hida, T., Blackshear, P.J., Klintworth, G.K., Swain, J.L., 1988, Tumor promoting phorbol esters induce angiogenesis *in vivo*. *Am. J. Physiol.* 23:C318-C322.

Müller-Decker, K., 1989, Interruption of TPA-induced signals by an antiviral and antitumoral xanthate compound: Inhibition of a phospholipase C type reaction. *Biochem. Biophys. Res. Commun.* 162:198-205.

Niedel, J.E., Kuhn, L.J., and Vanderbank, G.R., 1983, Phorbol diester receptor copurifies with protein kinase C. *Proc. Natl. Acad. Sci., USA*, 80: 36-40.

Rifkin, D.B., Gross, D., Moscatelli, D., and Gabrieledes, C., 1981, The involvement of poteases and protease inhibitors in neovascularization. *Acta Biol. Med. Cer.* 40:1259-1263.

Tsopanoglou, N.E., Pipili-Synetos, E., and Maragoudakis, M.E. Thrombin promotes angiogenesis by a mechanism independent of fibrin formation *Am. J. Physiology* (in press).

IV. INFLAMMATION/ATHEROSCLEROSIS

THE CLINICAL PROFILE OF ATHEROSCLEROSIS

Allan D. Callow

Department of Surgery
Washington University
St. Louis, MO 63110
U.S.A.

ATHEROSCLEROSIS

Current thinking supports the belief that underlying the onset of the atherosclerotic lesion is chronic injury to the arterial endothelium. The action of shear stress as a consequence of patterns of blood flow, whether turbulent or laminar, and as engendered by branch points and flow dividers, may be the origin of the injury. A number of factors such as hypercholesterolemia, cigarette smoking, hypertension and immune complexes may act synergistically with chronic endothelial injury to increase the degree and rate of plaque formation. That the response to these risk factors, such as flow disturbances, may be influenced by variations of susceptibility of endothelial cells is suggested by the great variability of the age at onset, rate of progression among individuals, and extent of plaque size in different anatomical beds. The predisposition to the development of symptomatic atherosclerotic lesions of the extracranial carotid artery, the coronary circulation, particularly at branch points, the infrarenal aorta, the iliofemoral tract of the lower extremities, and the renal arteries, particularly their ostia, is well known. By contrast, the vessels of the upper extremity, the mid descending thoracic aorta, most of the visceral vessels, with the occasional exception of the orifices of the celiac axis and the superior mesenteric artery are rarely the site of atherosclerotic disease and even more rarely productive of symptoms. This variation in vulnerability from arterial bed to arterial bed supports the concept of heterogeneity of endothelial response.

Lipid accumulation, both in macrophages as foam cells and as extracellular deposits, and a number of growth factors generated from macrophages, platelets and endothelium, initiate a phase of plaque formation characterized also by migration and proliferation of smooth muscle cells. Ultimately most, if not all, of such fibrolipid lesions acquire a fibrous and smooth muscle cap, shoulders of collagen, and a central core or cores of lipid. What governs the progression of the early atherosclerotic lesion to a large bulky plaque and, in turn, to a symptomatic compound plaque is largely unknown, except in general terms of the accepted risk factors. Thrombus formation, the result of hematoma development within the plaque or fissuring of the lumenal surface and deeper portions of the plaque appear to be major factors in plaque growth. Plaques, with fissuring, intraplaque hematoma, surface erosion and thrombus, are known as complicated or compound plaques. They may rapidly produce symptoms. Evidence strongly favors rapid progression once the atherosclerotic plaque has moved into the complicated phase. Thus, mural thrombosis and its progressive fibrous organization become major contributors to plaque progression in its mid to late stages of development (*Ross R, 1986; Glagov, et al., 1988; Richardson, et al., 1989; Fuster, et al., 1990; Thyberg, et al., 1990; Schwartz, et al., 1991*).

Vascular Endothelium, Edited by J.D. Catravas *et al.*
Plenum Press, New York, 1993

FIGURE 1: Behavioral Change Resulting From Cerebrovascular Disease. In February of 1945 near the end of World War II, Franklin Delano Roosevelt, Winston Churchill, and Joseph Stalin convened at Yalta for a decision-making meeting. The ill effects of their agreement have lasted until today and have contributed to conflicts in Korea, Vietnam, Czechoslovakia, and Poland. At that time, they all had advanced cerebrovascular disease. Roosevelt would die 3 months later from a massive brain hemorrhage, Stalin later from the same cause. Churchill had already had a series of small strokes, which would eventually leave him demented. Lord Charles Moran, Churchill's personal physician, described Roosevelt's appearance: He had a cape or shawl over his shoulders and appeared shrunken. He sat looking straight ahead with his mouth open, as if he were not taking things in. Everyone was shocked by his appearance and seemed to agree that the President had gone to bits physically. It was not only his physical deterioration that caught their attention. He intervened very little in the discussions, sitting with his mouth open. If he has sometimes in the past, been short on facts about the subject under discussion, his shrewdness has usually covered this up. Now, they say, the shrewdness has gone and there is nothing left.

THE HUMAN LESION

The human atherosclerotic plaque can be said to have its origin in dietary fat, the ultimate source of the lipoproteins of the liver. Characteristic features of the mature human plaque are the foam cells derived from macrophages, and to a lesser extent, from smooth muscle cells, the monocyte/macrophage and T-lymphocyte accumulations, presumably as inflammatory infiltrates, smooth muscle cell migration and proliferation, and subintimal lipid accumulation (*Faggiotto, et al., 1984; Gown, et al., 1986*). Steinberg and associates demonstrated that cultured macrophages are converted into foam cells by oxidized LDL cholesterol found in the human plaque and further demonstrated that LDL oxidation is necessary before it can lead to the formation of the plaque (*Steinberg, et al., 1989; Steinbrecher, et al., 1984*). Another and major characteristic of the human lesion is proliferation of smooth muscle cells following their phenotype conversion from the contractile to the synthetic phase.

Accumulation of extracellular matrix increases the bulk of the lesion, thereby reducing the lumen. This cellular proliferation of a fibrous cap over the surface of the plaque is one of the distinguishing features of the human from almost all animal models of atherosclerosis. The human atherosclerotic lesion is truly heterogeneous with this myriad of cell types and products.

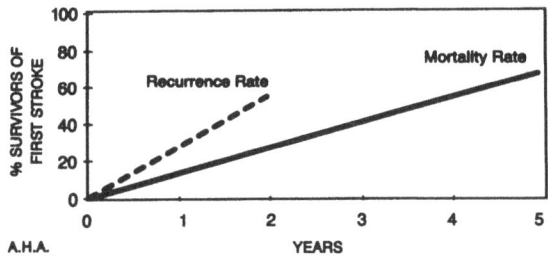

PROGNOSIS IN COMPLETED STROKES
(Cartoid Arterial System)
A. Initial Mortality for First Stroke - 20%
B. Mortality and Recurrence Rates Among
Survivors of First Stroke

FIGURE 2: Courtesy of the American Heart Association

THE CAROTID PLAQUE AND STROKE

It is estimated that any one time in the United States there are 2 million surviving stroke victims, a prevalence rate of roughly 9.7 per 100,000 persons. Stroke incidence as reported in 1976 was 137 per 100,000, a decline of the age-adjusted stroke incidence of 151 per 100,000 five years earlier. Stroke mortality had fallen considerably but was still 35.8 per 100,000 as reported in 1982 (*McDowell, 1985*). Despite this decline in stroke incidence and stroke mortality, approximately 8% of all deaths reported in the United States are directly due to stroke, causing stroke to rank as the fourth cause of death in this country. Among stroke victims who survive 30 days, two thirds are permanently disabled to some degree. The continuing risk for recurrent stroke among survivors is ominous with 21% of survivors suffering a second stroke, 7% a third stroke. Of 30 day survivors, two thirds are permanently disabled, one half survive at least five years, one third require prolonged inpatient rehabilitation (*McDowell, 1985*). Initial stroke incidence increases with age. Per 100,000 population, annual average initial stroke incidence is 341.6 from 55 to 64, 658 at ages 65-74, and at 75 to 84 it reaches 1,713. These figures apply to the male sex, but the female does not fare much better.

The symptomatic carotid plaque is an advanced lesion with a fibrous cap, a central atheromatous core of semi-liquid and crystalline cholesterol, and cellular debris plus a shoulder of smooth muscle cells, foam cells and a few T lymphocytes. Unlike the symptomatic lesions developing in the coronary circulation and the aortoilio-femoral arterial bed, where the intrusion of the plaque into the lumen causes symptoms by limiting the response to the call for increased blood flow during exercise, the carotid lesion usually becomes symptomatic by embolization following loss of integrity of the surface layer. Hemorrhage may occur within the plaque, possibly from vasa vasorum and equally possibly from the neoangiogenesis often surrounding the atherosclerotic plaque. An intraplaque, intramural hematoma forms. Progression to liquefaction and cell death plus erosion of the overlying cap permits extrusion of plaque contents. Cholesterol crystals, cellular debris and portions of clot, platelets and fibrin comprise the emboli. Conventional wisdom holds that these emboli travel in the distal arterial stream until they come to lodge in a branch vessel within the brain so small as to prevent further passage. Depending upon the cerebral territory in which the embolus lodges, the signs and symptoms may cover the entire spectrum of cerebral function, or if in a *silent* area, be entirely asymptomatic. A very small embolus may produce temporary neuronal malfunction with symptoms lasting only a few seconds to a few minutes. The embolus, particularly if it is of platelet composition may break up, move on and become dispersed. Larger emboli or showers of small emboli may

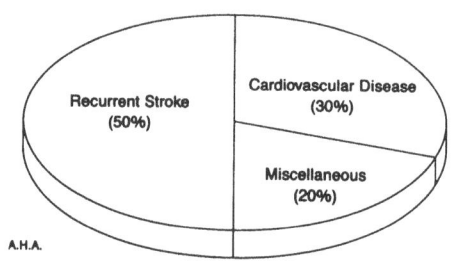

PROGNOSIS IN COMPLETED STROKES
(Cartoid Arterial System)
C. Eventual Causes of Death After First Stroke

FIGURE 3: Courtesy of the American Heart Association

produce permanent damage with a fixed neurologic deficit, the type of which, that is motor or sensory or both, and the extent of which, depend upon the area of lodgement. What are called *pressure significant* or *hemodynamically significant* internal carotid artery lesions are rarely symptomatic because of the richness of the cervical and cerebral collateral circulation. The normal Circle of Willis provides for easy access of blood from the posterior circulation, via the vertebrobasilar, and posterior cerebral branches to the anterior circulation and from left to right hemisphere (or vice versa) via the internal to the middle and anterior cerebral vessels. Thus, either the offending occlusive lesion must be severely stenotic or actually occlusive or accompanied by severe contralateral occlusive disease as well, or the collateral routes markedly deficient to result in symptoms due to flow reduction.

In short, the signs and symptoms of cerebral occlusive disease of the extracranial circulation are based upon emboli from a far advanced and degenerating atherosclerotic plaque rather than from gradual, steady reduction of flow. The clinical manifestation of disease, motor or sensory loss of function, occurs at a distance from the location of the plaque, most often at the bifurcation of the common into the internal and external carotid arteries. The blood vessels which are the target of the disease, namely the middle cerebral and other intracranial branches usually show little or no atherosclerotic disease themselves.

Thrombus may form upon the surface of an ulcerated plaque from blood within the lumen as a consequence of deposition of fibrin and blood components. Such surface clots may also give rise to emboli and the relative frequencies of each mechanism, intraplaque hematoma dissolution or luminal surface thrombus formation are unknown. The embolic etiology of stroke may explain why antiplatelet and anticoagulant agents are less effective in reducing stroke incidence than that of myocardial infarction. Plaque growth is unpredictable, non-linear, and probably episodic. This observation is based in part on the noted sudden increase in size of plaque and diminution of lumen, presumably due to thrombus formation on the plaque, hematoma formation in the plaque or both. A major but poorly understood stage in plaque growth is plaque fissuring. Its pathogenesis and the forces governing its extent are unknown. Once formed, blood enters the plaque, forms a stable thrombus, and undergoes organization with collagen accumulation and smooth muscle cell proliferation. Healing of the fissure occurs and, for an unknown time the patient is protected from further embolization. Autopsy material has provided evidence of healed fissures. If fissuring and intraplaque hematoma formation can *heal*, the lesion may stabilize and become asymptomatic. Calcification of the hematoma may occur. Cessation of transient ischemic attacks may also occur after an occluding thrombus closes the lumen. Such occlusions may be asymptomatic. The fundamental difference is that whereas stenosing and occluding lesions in other arterial beds, particularly the coronary circulation and the vessels of the lower extremity, produce symptoms chiefly by reduction of flow produced by the insidious enlargement of the hemodynamically

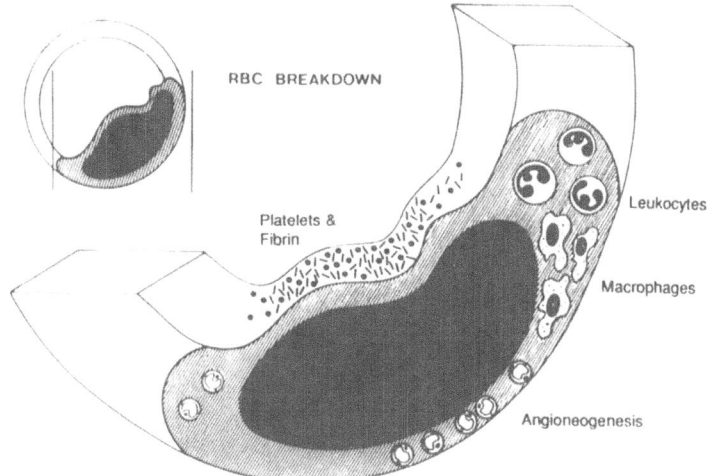

RBC BREAKDOWN

Platelets & Fibrin

Leukocytes

Macrophages

Angioneogenesis

FIGURE 4: Diagram illustrating inflammatory response to an intraplaque hemorrhage with leukocyte and macrophage migration occurring as well as angioneogenesis. Platelets and fibrin are shown depositing on the luminal surface over the hemorrhage. (*Lusby, Ferrel, Wylie*).

significant lesion resulting in under perfusion of cardiac or skeletal muscle, particularly obvious with increased exercise and the accompanying increased demand for blood, the carotid lesion produces symptoms chiefly by embolization. Although cerebral symptoms can be also the result of reduced flow rather than emboli, this appears to be true only in the presence of advanced lesions, with limitation of collateral flow on a congenital basis, or contralateral occlusive disease as stated earlier.

CORONARY PLAQUE AND MYOCARDIAL INFARCTION

Development of the atherosclerotic lesion appears to be governed by the same factors and to progress in the same stepwise fashion irrespective of the arterial bed. Thus, in the carotid arteries and the aortoiliac vessels the first event is early injury of the endothelium, presumably due to shear stress factors potentiated by hypercholesterolemia, cigarette smoking and hypertension. Following damage to the endothelium, there is transendothelial migration of monocyte/macrophages, lipid accumulation, formation of foam cells, accumulation of collagen, fibrous material, and T-lymphocytes. Although this typical lesion is large enough to intrude upon the lumen of the vessel, not until intrusion reaches 50 to 60% of the internal diameter of the artery will there be detectable reduction of blood flow or perfusion pressure. Indeed, the luminal narrowing may reach 75% with no reduction in flow because of compensatory dilatation of the artery which accompanies the atherosclerotic process. This is particularly true in the coronary circulation. Progressive flow reduction results in a more pronounced disparity between coronary artery oxygen delivery and myocardial oxygen need. Clinically this is manifested by the onset of exertional angina. Attacks of angina become more frequent, more severe, and precipitated by less exertion as coronary flow reduction increases with the enlarging plaque. Erosion of the surface and disruption of the plaque occur just as in the carotid plaque. However, a number of observations suggest substantial differences in the mechanisms by which the disrupted carotid and coronary plaques give rise to symptoms. As noted earlier, the contents of the carotid plaque are extruded in the form of multiple emboli which travel to the distal cerebral circulation, lodge in a vessel too small to accommodate their further passage, and lead to malfunction or infarction of the target neurons. Disruption of the coronary plaque appears to be followed by repeated episodes of mural thrombus formation and fibrous organization. Layered thrombi overlying fissured plaques have

93

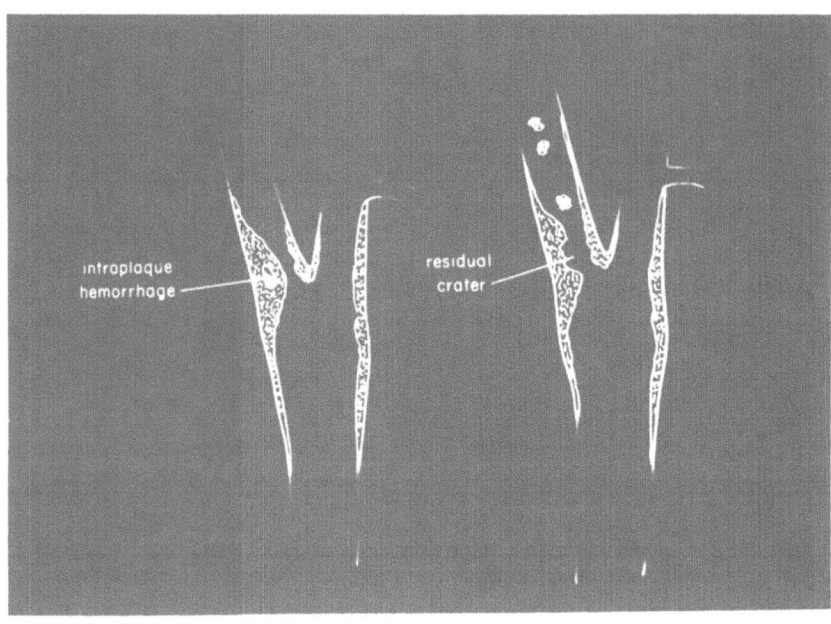

FIGURE 5: Ulceration and embolization schematically depicted at the carotid bifurcation.

been reported (*Falk, E., 1985*) suggesting that recurrent mural thrombus formation may be the basis for gradual vascular occlusion. When acute and complete, such thrombus formation leads to myocardial infarction or unstable angina. Sudden death is a constant threat. Platelet and fibrin degradation products have been detected in the compound atherosclerotic plaque further supporting the concept that thrombus formation and its organization are associated with plaque growth (*Bini, et al., 1989; Smith, et al., 1990; Wilcox, et al., 1988; Bruschke, et al., 1989*). As in other arterial beds, evidence of plaque emboli has been found in the small branches of the coronary circulation, but the major factor leading to onset of symptoms in the coronary circulation appears to be repetitive mural thrombus formation and organization resulting in acute or gradual arterial obstruction.

Coronary vasoconstriction has been observed at angiography during and after percutaneous transluminal coronary angioplasty and the plaque disruption and thrombosis that accompanies it (*Hackett et al., 1987; Fischell et al., 1988; el-Tamini et al., 1991*). The endothelium plays a major role in control of vasomotor tone through such mediators as prostacyclin, endothelium-derived relaxing factor (EDRF) and the constrictor peptide endothelin (*Fischell et al., 1988; el-Tamini et al., 1991; Maseri et al., 1978; Vanhoutte and Shimokawa, 1989; Vane et al., 1990; Moncada et al., 1976; Furchgott and Szwadzki, 1980; Palmer et al., 1988; Yanagisawa et al., 1988; Endothelins, 1991*). The atherosclerotic coronary artery develops a constrictor response when exposed to acetylcholine in humans as well as in experimental animals (*Yeung et al., 1991; Cohen and Zitnay, 1988*). Diminution of EDRF secretion in the human atherosclerotic coronary has been reported (*Chester et al., 1990*).

Spasm has also been observed in association with the atherosclerotic occlusive vessels of the lower extremities: the iliac, femoral, and popliteal and distal vessels below. The major cause of symptoms, however, is due to gradual encroachment on the lumen and reduction of flow by the offending plaque. For a long and indeterminate period of time, the plaque may remain entirely asymptomatic. When demand for oxygen increases, as with exercise, blood supply is insufficient. Vasodilatation may be unable to compensate. Pain in the muscles of the lower extremity occurs with oxygen lack as it does in the myocardium with coronary insufficiency. With rest and rapid restoration of the normal balance between blood supply and demand, pain diminishes and disappears. On the other hand, atherosclerosis is a progressive disease. As the lesion enlarges further, there may be not only pain with exercise, but also pain at rest.

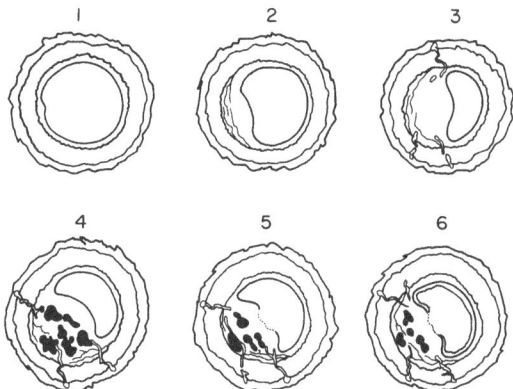

FIGURE 6: Schematic representation of progression of an atherosclerotic lesion: 1) a normal artery; 2) subintimal lipid accumulation; 3) beginning neoangiogenesis; 4) intraplaque hemorrhage; 5) endothelial loss and 6) terminal thrombotic occlusion.

The intermittent pain of exercise is known as intermittent claudication. The term, from the Latin, literally means intermittent *limping* from *claudicare*. Pain at rest indicates severe ischemia and is usually accompanied by skin and muscle mass wasting. Frequently it is a harbinger of gangrene.

REGRESSION

Reduction of plaque size following intensive efforts to lower cholesterol levels in patients at high risk has been reported with increasing frequency. These studies, for the most part, are based upon repeated angiographic estimation of the lesion size as denoted by lumen diameter seen on patient entry into the study and one or more years later after cholesterol lowering efforts (*Brown et al., 1990; Buchwald et al., 1990; Blankenhorn et al., 1988; Kane et al., 1990*). A variety of cholesterol lowering agents have been administered to patients for periods of 2 to 5 years. These consisted of gemfibrozil, cholestipol, niacin, lovastatin, together with dietary restrictions (*Brown et al., 1990; Frick et al., 1987; Blankenhorn et al., 1990; Ornish et al., 1990*). In addition to modest increases in HDL cholesterol levels and slightly greater decreases in LDL, cholesterol and triglycerides levels, the lowered incidence of cardiac end points was impressive (*Brown et al., 1990; Buchwald et al., 1990*). However, the alleged changes in luminal diameter seen in the repeat arteriograms are small at best and are difficult to reconcile with the more impressive reduction in cardiac end points suggesting that the reduction in cardiac end points is due to reduction in the progression rate of lesions. Evidence favors the belief that these measures halt progression of the lesion more than causing actual reduction of plaque size.

Considerable confusion continues concerning the protective effect of elevation of the HDL level, reduction of the LDL level, a combination of elevation of HDL and lowering of LDL, and the mechanisms by which these changes generate their protective effects. One attractive hypothesis is that *reverse cholesterol transport* may be engendered by HDL (*Reichl and Miller, 1989*). Rabbits receiving homologous HDL cholesterol intravenously experienced inhibition of fatty deposition in the aortic wall and diminution of the lesions already present despite a high intake of dietary cholesterol (*Reichl and Miller, 1989; Badimon et al., 1990*). Transgenic mice have shown an increase in HDL cholesterol levels following overexpression of the human apolipoprotein A-I gene with some evidence of an antiatherogenic effect (*Rubin et al., 1991*). Inasmuch as LDL must undergo oxidation to enter the cell, inhibition of the process by the use of antioxidants deserves continuing investigation (*Steinberg, 1991*).

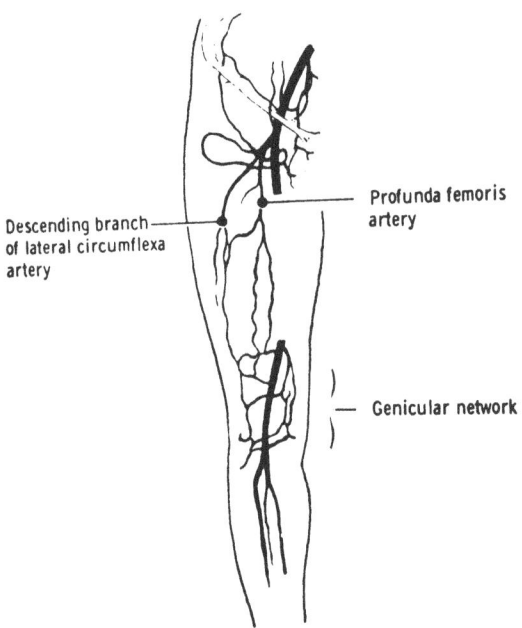

Descending branch
of lateral circumflexa
artery

Profunda femoris
artery

Genicular network

FIGURE 7: Diagram of arterial supply to the lower extremity showing 1) a frequent pattern of occlusion of the main channel (the superficial femoral artery) and 2) collateral vessels around the block.

The beneficial effect of antiplatelet agents and anticoagulants in reducing the incidence of myocardial infarction have not been duplicated in carotid stroke (Antiplatelet Trialists, 1988; *Hennekens et al., 1989; Smith, 1992*). Some understanding of this disparity may be gained by recalling the different mechanisms by which symptoms are produced in the cerebral and coronary circulations. The carotid plaque undergoes necrosis with extrusion of contents and embolization to intracranial vessels. Antiplatelet agents and anticoagulants cannot control these events. Thrombus formation, either on the surface of the plaque or within the center of the carotid plaque, most certainly occurs, but it is usually followed by fragmentation of the surface clot or of the contents of the plaque. Some layering of clot with organization, and fibrous conversion with stabilization, may also occur, and the carotid may undergo occlusion. By contrast, thrombus formation in or on the coronary plaque usually undergoes organization with stimulation of cellular proliferation and the gradual layered buildup of the lesion until occlusion occurs. This may be the terminal event producing myocardial infarction and even death. The collateral circulation available in the coronary circulation is contained within the heart. Unlike the brain, there are no extrinsic or extracardiac sources. Interfering with platelet activation by aspirin or other antiplatelet agents may thus inhibit the initial phase of clot formation.

REFERENCES

Antiplatelet Trialists; Collaboration, 1988, Secondary prevention of vascular disease by prolonged anti-platelet treatment. *BMJ* 296:320-31.

Badimon, J.J., Badimon, L., Fuster, V., 1990, Regression of atherosclerotic lesions by high density lipoprotein plasma fraction in the cholesterol-fed rabbit. *J. Clin. Invest.* 85:1234-41.

Bini, A., Fenoglia, J.J. Jr., Mesa-Tejada, R., Kudryk, B., Kaplan, K.L., 1989, Identification and distribution of fibrinogen, fibrin, and fibrin(ogen) degradation products in atherosclerosis: use of monoclonal antibody. *Arteriosclerosis* 9:109-121.

Blankenhorn, D.H., Johnson, R.L., Mack, W.J., el Zein, H.A., Vailas, L.I., 1990, The influence of diet on the appearance of new lesions in human coronary arteries. *JAMA* 263:1646-52.

Blankenhorn, D.H., Nessim, S.A. Johnson, R.L., Sanmarco, M.E., Azen, S.P. Cashen-Hemphill, L., 1988, Beneficial effects of combined colestipol-niacin therapy on coronary atherosclerosis and coronary venous bypass grafts. *JAMA* 259:2698.

Brown, G., Albers, J.J., Fisher, L.D., et al., 1990, Regression of coronary artery disease as a result of intensive lipid-lowering therapy in men with high levels of apolipoprotein B. *N. Engl. J. Med.* 323:1289-98.

Bruschke, A.V.G., Kramer, J.R. Jr., Bal, T.E., Haque, I.U., Detrano, R.C., Goormastic, M., 1989, The dynamics of progressive coronary atherosclerosis studied in 168 medically treated patients who underwent coronary angiography three times. *Am. Heart J.* 117:296-305.

Buchwald, H., Varco, R.L., Matts, J.P., et al., 1990, Effect of partial ileal bypass surgery on mortality and morbidity from coronary heart disease in patients with hypercholesterolemia: report of the Program on the Surgical Control of the Hyperlipidemias (POSCH). *N. Engl. J. Med.* 323:946-55.

Chester, A.H., O'Neill, G.S., Moncada, S., Tadjkarimi, S., Yacoub, M.H., 1990, Low basal and stimulated release of nitric oxide in atherosclerotic epicardial coronary arteries. *Lancet* 336:897-900.

Cohen RA, Zitnay, K.M., Haudenschild, C.C., Cunningham, L.D., 1988, Loss of selective endothelial cell vasoactive stimuli in hypercholesterolemia in pig coronary arteries. *Circ. Res.* 63:903-10.

el-Tamini, H., Davies, G.J., Hackett, D., et al., 1991, Abnormal vasomotor changes early after coronary angioplasty. *Circulation* 84:1198-202.

Endothelins, 1991, *Lancet* 337:79-81.

Faggiotto, A., Ross, R., Harker, L., 1984, Studies of hypercholesterolemia in the non-human primate. Changes that lead to fatty streak formation. *Arteriosclerosis* 4:323-330.

Falk, E., 1985, Unstable angina with fatal outcome: dynamic coronary thrombosis leading to infarction and/or sudden death: autopsy evidence of recurrent mural thrombosis with peripheral embolization culminating in total vascular occlusion. *Circulation* 71:699-708.

Fischell, T.A., Derby, G., Tse, T.M., Stadius, M.L., 1988, Coronary artery vasoconstriction routinely occurs after percutaneous transluminal coronary angioplasty. *Circulation* 78:1323-4.

Frick, M.H., Elo, O., Haapa, K., et al., 1987, Helsinki Heart Study: primary prevention trial with gemfibrozil in middle-aged men with dyslipidemia: safety of treatment, changes in risk factors, and incidence of coronary heart disease. *N. Engl. J. Med.* 317:1237-45.

Furchgott, R.F., Zawadzki, J.V., 1980, The obligatory role of endothelial cells in the relaxation of arterial smooth muscle by acetylcholine. *Nature* 299:373-6.

Fuster, V., Stein, B., Ambrose, J.A., Badimon, L., Badimon, H., Chesebro, J.H., 1990, Atherosclerotic plaque rupture and thrombosis: evolving concepts. *Circulation* 82 Suppl II:II-47-II-59.

Glagov, S., Zarins, C., Giddens, D.O., Ku, D.N., 1988, Hemodynamics and atherosclerosis: insights and perspectives gained from studies of human arteries. *Arch. Pathol. Lab. Med.* 112:1018-1031.

Gown, A.M., Tsukada, T., Ross, R., 1986, Human atherosclerosis II. Immunocytochemical analysis of the cellular composition of human atherosclerotic lesions. *Am. J. Path.* 125:191-207.

Hackett, D., Davies, G., Choierchia, S., Maseri, A., 1987, Intermittent coronary occlusion in acute myocardial infarction: value of combined thrombolytic and vasodilatory therapy. *N. Engl. J. Med.* 2317:1055-9.

Hennekens, C.H., Buring, J.E., Sandecock, P., Collins, R., Peto, R., 1989, Aspirin and other antiplatelet agents in the secondary and primary prevention of cardiovascular disease. *Circulation* 80:-749-56.

Kane, J.P., Malloy, M.K.J., Ports, T.A., Phillips, N.R., Diehl, J.C., Havel, R.J., 1990, Regression of coronary atherosclerosis during treatment of familial hypercholesterolemia with combined drug regimens. *JAMA* 264:3007-12.

Maseri, A., L´Abbate, A., Baroldi, G., 1978, Coronary vasospasm as a possible cause of myocardial infarction: a conclusion derived from the study of "preinfarction" angina. *N. Engl. J. Med.* 299:1271.

McDowell, F.H., Caplan, L.R. (eds.). Cerebrovascular Survey Report, 1985, Bethesda, MD for the National Institute of Neurological and Communicative Disorders and Stroke, NIH, US Public Health Service.

Moncada, S., Gryglewski, R., Bunting, S., Vane, J.R., 1976, An enzyme isolated from arteries transforms prostaglandin endoperoxides to an unstable substance that inhibits platelet aggregation. *Nature* 263:663-5.

Ornish, D., Brown, S.E., Scherwitz, L.W., et al., 1990, Can lifestyle changes reverse coronary heart disease? The Lifestyle Heart Trial. *Lancet* 336:129-33.

Palmer, R.M.J., Ashton, D.S., Moncada, S., 1988, Vascular endothelial cells synthesize nitric oxide from L arginine. *Nature* 333:664-6.

Reichl, D., Miller, N.E., 1989, Pathophysiology of reverse cholesterol transport: insights from inherited disorders of lipoprotein metabolism. *Arteriosclerosis* 9:785-97.

Richardson, P.D., Davies, M.J., Born, G.V.R., 1989, Influence of plaque configuration and stress distribution on fissuring of coronary atherosclerotic plaques. *Lancet* 2:941-4.

Ross, R., 1986, The pathogenesis of atherosclerosis - an update. *N. Engl. J. Med.* 314:488-500.

Rubin, E.M., Krauss, R.M., Spangler, E.A., Verstuyft, J.G., Clift, S.M., 1991, Inhibition of early atherogenesis in transgenic mice by human apolipoprotein AI. *Nature* 353:265-7.

Schwartz, C.J., Valente, A.J., Sprague, E.A., Kelley, J.L., Nerem, R.M., 1991, The pathogenesis of atherosclerosis. *Clin. Cardiol.* 14 Suppl I:I-1-I-16.

Smith, P., 1992, Antithrombotic therapy in the chronic phase of myocardial infarction. In: Fuster, V., Verstraete, M., eds. *Thrombosis in Cardiovascular Disorders* Philadelphia: E.B. Saunders, 529-44.

Smith, E.B., Kean, A., Grant, A., Stirk, C., 1990, Fate of fibrinogen in human arterial intima. *Arteriosclerosis* 10:263-275.

Steinberg, D., 1991, Antioxidants and atherosclerosis. *Circulation* 84:1420-5.

Steinberg, D., Partharsayrathy, S., Carew, T.E., Kohoo, J.C., and Witztum, J.L., 1989, Beyond cholesterol:modification of low density lipoprotein that increases its atherogenecity. *N. Engl. J. Med.* 320:915-924.

Steinbrecher, U.P., Partharsayrathy, S., Leake, D.S., Witztum, J.L., and Steinberg, D., 1984, Modification of low density lipoprotein by endothelial cells involves lipid peroxidation and degradation of low density lipoprotein phospholipids. *Proc. Natl. Acad. Sci., USA* 83:3883-3887.

Thyberg, J., Hedin, U., Sjolund, M., Palmberg, L., Bottger, B.A., 1990, Regulation of differentiated properties and proliferation of arterial smooth muscle cells. *Arteriosclerosis* 10:966-90.

Vane, J.R., Anggaard, E.E., Botting, R.M., 1990, Regulatory functions of the vascular endothelium. *N. Engl. J. Med.* 323:27036.

Vanhoutte, P.M., Shimokawa, H., 1989, Endothelium-derived relaxing factor and coronary vasospasm. *Circulation* 80:1-9.

Wilcox, J.N., Smith, K.M., Williams, L.T., Schwartz, S.M., Gordon, D., 1988, Platelet-derived growth factor mRNA detection in human atherosclerotic plaques by *in situ* hybridization. *J. Clin. Invest.* 82:1134-43.

Yanagisawa, M., Kurihara, H., Kimura, S., et al., 1988, A novel potent vasoconstrictor peptide produced by vascular endothelial cells. *Nature* 332:411-5.

Yeung, A.C., Vekshtein, V.I., Krantz, D.S., et al., 1991, The effect of atherosclerosis on the vasomotor response of coronary arteries to mental stress. *N. Engl. J. Med.* 325:1551-1556.

PROINFLAMMATORY ROLE OF LEUKOCYTE ADHESION MOLECULES

*John M. Harlan, **'****Robert K. Winn, and **Nicholas B. Vedder

*Department of Medicine/Hematology
University of Washington
Seattle, WA 98195
U.S.A.

**Department of Surgery
University of Washington
Seattle, WA 98195
U.S.A.

***Department of Biophysics
University of Washington
Seattle, WA 98195
U.S.A.

INTRODUCTION

The adhesive interaction between circulating leukocytes and the vessel wall is a critical event in host defense against microbial invasion and in the repair of tissue damage. Altered leukocyte-endothelial interactions, however, may sometimes contribute to the pathogenesis of vascular and tissue injury in inflammatory and immune disorders. This chapter will briefly summarize our current understanding of the molecular basis of leukocyte adherence to endothelium in inflammatory and immune reactions and discuss *anti-adhesion* therapy as a novel approach to the treatment of a wide spectrum of clinical disorders.

LEUKOCYTE-ENDOTHELIAL ADHESION MOLECULES

Studies utilizing intravital microscopy of vessels in the microcirculation have identified a sequence of events involved in the emigration of leukocytes from the bloodstream to extravascular sites of inflammation or immune reaction (Figure 1, *Harlan et al., 1992*). In response to extravascular stimuli, signals are generated which activate both the leukocyte and the endothelial cell. Activation leads initially to transient leukocyte adherence to endothelium which is manifested by "rolling" along the vessel wall. Atherton and Born ascribed rolling to the interaction of shear and adhesive forces (*Atherton and Born, 1973*): "*As a result of biochemical changes effected in the walls by inflammatory agents, granulocytes colliding with the walls experience an adhesive force as well as the shear force exhibited at the wall by the blood flow. The rolling movement is then the resultant of these two forces.*"

Rolling is often followed by sustained adhesion or "sticking" of leukocytes to the endothelium and then subsequent diapedesis. Leukocyte sticking to the vessel wall at sites of inflammation was recognized as a hallmark of the inflammatory response by

Vascular Endothelium, Edited by J.D. Catravas *et al.*
Plenum Press, New York, 1993

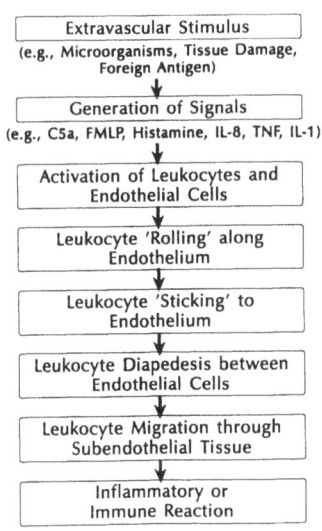

```
        ┌─────────────────────────────┐
        │    Extravascular Stimulus    │
        └─────────────────────────────┘
       (e.g., Microorganisms, Tissue Damage,
                 Foreign Antigen)
                      ↓
        ┌─────────────────────────────┐
        │      Generation of Signals   │
        └─────────────────────────────┘
       (e.g., C5a, FMLP, Histamine, IL-8, TNF, IL-1)
                      ↓
        ┌─────────────────────────────┐
        │   Activation of Leukocytes and │
        │        Endothelial Cells     │
        └─────────────────────────────┘
                      ↓
        ┌─────────────────────────────┐
        │   Leukocyte 'Rolling' along  │
        │         Endothelium          │
        └─────────────────────────────┘
                      ↓
        ┌─────────────────────────────┐
        │   Leukocyte 'Sticking' to    │
        │         Endothelium          │
        └─────────────────────────────┘
                      ↓
        ┌─────────────────────────────┐
        │  Leukocyte Diapedesis between │
        │       Endothelial Cells      │
        └─────────────────────────────┘
                      ↓
        ┌─────────────────────────────┐
        │   Leukocyte Migration through │
        │      Subendothelial Tissue   │
        └─────────────────────────────┘
                      ↓
        ┌─────────────────────────────┐
        │      Inflammatory or         │
        │       Immune Reaction        │
        └─────────────────────────────┘
```

FIGURE 1: Sequences of events involved in leukocyte emigration.

the earliest intravital microscopists and was described in great detail by later investigators, particularly Clark and Clark (1935): *"(A leukocyte) could be seen to glide smoothly along or roll over and over along the wall, depending on the speed of the circulation at time, until it arrived at a particular point, when it stopped and remained adherent to the wall for a few seconds or several minutes."*

It was a matter of some debate whether alterations in the leukocyte or the endothelial cell were responsible for these adhesive interactions. Cohnheim (1989) emphasized molecular alterations in the vessel wall as the cause of leukocyte sticking, whereas, Metchinikoff (1893) stressed the importance of chemotactic influences upon the leukocyte (*Movat, 1989*). It was not until the past decade that the molecular basis of these adhesive interactions was defined. Perhaps not surprising, it is now apparent that both the leukocyte and the endothelial cell are involved. Table 1 lists the leukocyte and endothelial cell adhesion molecules that have been identified to date. They comprise two categories of receptor-ligand interactions: selectin receptors expressed on both leukocytes and endothelial cells that recognize specific carbohydrate counter-structures, and leukocyte integrin receptors that bind to ligands on the endothelial cells that are members of the immunoglobulin supergene family. Figure 2 illustrates our current understanding of the contribution of these molecules to the adhesive interaction of leukocytes with endothelial cells. Studies *in vitro* (*Lawrence and Springer, 1991*) and *in vivo* (*Ley et al., 1991; von Adrian et al., 1991*) indicate that selectin receptors and their carbohydrate counter-structures mediate leukocyte rolling. In contrast, sticking and diapedesis are dependent upon the interaction of leukocyte integrins with endothelial ligands (*Harlan et al., 1985; Smith et al., 1989*).

Activation events play an important role in regulating leukocyte-endothelial interactions (*Butcher, 1991; Figdor et al., 1992, Zimmerman et al., 1992*). For example, stimulation of neutrophils by a chemoattractant produces a transient increase in the avidity of L-selectin followed rapidly by its shedding from the cell surface, an increase surface expression of CD11b/CD18, and an increase in avidity of CD11a/CD18 and CD11b/CD18. Similarly, activation of lymphocytes produces a shedding of L-selectin, and an increase in avidity of CD11a/CD18 and VLA-4. Endothelial activation by thrombin, histamine, LTC4/LTD4 and hydrogen peroxide results in the translocation of P-selectin from intracellular granules to the surface membrane. ICAM-1 is induced by interferon-gamma, tumor necrosis factor (TNF), interleukin-1 (IL-1), and lipopolysaccharide (LPS). Interleukin-4 increases surface expression of VCAM-1, and both E-selectin and VCAM-1 are induced by TNF, IL-1, and LPS.

TABLE 1: Leukocyte-Endothelial Cell Adhesion Molecules[+]

Adhesion Molecules	Leukocyte	Endothelial Cell
Integrin-Immunoglobulin[*]		
	CD11a/CD18 (LFA-1)	ICAM-1, ICAM-2
	CD11b/CD18 (Mac-1, Mo1, CR3)	ICAM-1, ? other
	CD11c/CD18 (p150, 95)	?
	VLA-4 (CD49d/CD29)	VCAM-1
Selectin-Carbohydrate[**]		
	L-selectin	?
	Sialylated Lewis X (? other)	E-selectin
	Sialylated Lewis X (? other)	P-selectin

[+]For reviews see: Butcher, 1991; Carlos and Harlan, 1990; Patarroyo et al; Springer, 1990; Zimmerman et al., 1992.
[*]Leukocyte integrin receptors bind to ligands on the endothelial cell that are members of the immunoglobulin supergene family.
[**]Selectin receptors expressed on both cell types recognize specific carbohydrate counter-structures.

The selective recruitment of subpopulations of leukocytes (e.g., neutrophils, monocytes, eosinophils, or memory T-lymphocytes) to sites of inflammation or immune reaction is determined both by the constitutive expression of adhesion molecules and by their stimulus-induced expression and activation (*Butcher, 1991*). For example, in many inflammatory settings, activation of endothelium likely results in the expression of both P- and E-selectins, thereby facilitating the initial transient adhesion of neutrophils. Subsequent firm sticking and diapedesis is mediated by the interaction of CD11b/CD18 and CD11a/CD18 with ICAM-1, initiated by activation of the beta-2 integrin receptors by endothelial surface-expressed platelet activating factor or locally-generated chemoattractants, such as interleukin-8, C5a, or leukotriene B4. Since neutrophils lack VLA-4, they will not bind when the inflammatory milieu is such that the endothelium at sites of inflammation expresses predominantly VCAM-1, whereas, under these circumstances mononuclear leukocytes will adhere and diapedese, if their integrin receptors are appropriately activated by chemoattractants (e.g., monocyte chemoattractant protein-1 for monocytes, interleukin-4 or -5 for eosinophils). The regulation of adhesion receptor-ligand interactions by a diverse array of activating stimuli, such as cytokines, lipids, complement components, or oxidants, is certain to be an exceedingly complex process.

ANTI-ADHESION THERAPY

The importance of leukocyte adherence to endothelium in host defense and repair is dramatically illustrated by Leukocyte Adhesion Deficiency (LAD), the syndrome resulting from deficiency of the CD11/CD18 beta-2 integrin complex (*Anderson and Springer, 1987*). Phagocytes from affected patients are unable to adhere to and migrate across endothelium at sites of inflammation resulting in a profound defect in phagocyte emigration. As a consequence, severely-affected patients suffer from recurrent, sometimes life-threatening, bacterial infections. Although deficient leukocyte-endothelial interactions can obviously be quite detrimental, exaggerated, or inappropriate leukocyte adherence to endothelium may also have deleterious consequences for the host. Altered leukocyte-endothelial interactions have been impli-

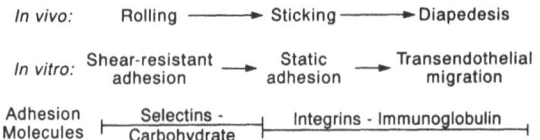

FIGURE 2: Adhesion molecules involved in leukocyte-endothelial interactions *in vitro* and *in vivo*.

cated in the pathogenesis of a wide variety of clinical disorders, some of which are listed in Table 2.

Leukocyte adherence to endothelium appears to be necessary for leukocyte-mediated vascular injury to occur (*Diener et al., 1985*). Close approximation of leukocyte to endothelium during adherence and transmigration produces a protected microenvironment at the cell-cell interface. Toxic products such as oxidants and proteases released by the adherent neutrophil at this site are inaccessible to oxygen radical scavengers and protease inhibitors. Inhibition of leukocyte adherence to endothelium-*anti-adhesion* therapy - prevents formation of the protected microenvironment and is therefore effective in reducing endothelial injury by activated leukocytes. Inhibition of leukocyte adherence to endothelium also prevents emigration to tissue, and consequently, reduces tissue damage produced by emigrated leukocytes. Finally, inhibition of leukocyte adherence to endothelium or leukocyte-leukocyte aggregate formation (homotypic adhesion) can prevent microvascular occlusion.

Leukocyte adhesion, especially neutrophil adhesion, may play a particularly important role in disorders associated with ischemia and reperfusion (*Hernandez et al., 1987*). It appears that in some settings an inflammatory stimulus(i) is generated at the time of reperfusion that activates leukocytes and/or endothelial cells. The identity of the activating agent(s) is uncertain but oxidants, leukotrienes, platelet activating factor, TNF and activated complement components have all been implicated in various models. The activation of leukocytes or endothelial cells during ischemia-reperfusion

TABLE 2: Clinical Disorders Possibly Associated with Leukocyte-Mediated Vascular and Tissue Injury

Ischemia-Reperfusion Syndromes	Inflammatory/Immune Disorders
Myocardial Infarction	Sepsis
Shock	Adult Respiratory Distress Syndrome
Stroke	Multiple Organ Failure Syndrome
Limb Replantation	Allograft Rejection
Organ Transplantation	Rheumatoid Arthritis
Cardiopulmonary Bypass	Atherosclerosis
	Inflammatory Bowel Disease
	Cerebral Edema in Meningitis
	Graft Versus Host Disease
	Multiple Sclerosis
	Vasculitis Syndromes
	Inflammatory Dermatoses
	Diabetes
	Asthma
	Aspiration Pneumonitis
	Burns
	Frostbite

TABLE 3: Anti-Adhesion Therapy in Experimental Models

	mAb	References
Ischemia-Reperfusion		
Heart	CD11b	*Simpson et al., 1990*
	CD18	*Ma et al., 1991*
Intestine	CD18	*Hernandez, 1987*
Lung	CD18	*Horgan et al., 1990*
		Bishop et al., 1992
Extremity	CD18	*Carden et al., 1990*
		Vedder et al., 1990
Hemorrhagic shock	CD18	*Vedder et al., 1988*
		Mileski et al., 1990
Central nervous system	CD18	*Clark et al., 1991a*
	CD54	*Clark et al., 1991b*
Inflammatory/Immune		
Meningitis	CD18	*Tuonamen et al., 1989*
		Saez-Llorens et al., 1991
Allograft rejection	CD54	*Cosimi et al., 1990*
Diabetes	CD11b	*Hutchings et al., 1990*
Asthma	CD54	*Wegner et al., 1990*
	E-selectin	*Gundel et al., 1991*
Shwartzman reaction	CD18, CD54	*Argenbright and Barton, 1992*
Acute lung injury	E-selectin	*Mulligan et al., 1991a*
	CD11b, CD18	*Mulligan et al., 1991b*
Arthritis	CD49d	*Issekutz and Issekutz, 1990*
Multiple sclerosis	CD49d	*Yednock et al., 1992*

results in leukocyte adhesion, producing microvascular occlusion or vascular injury. Inhibition of leukocyte adhesion to endothelium or leukocyte aggregation will reduce vascular and tissue injury in this setting.

Monoclonal antibodies directed to functional epitopes on leukocyte or endothelial cell adhesion proteins have provided powerful tools to assess the contribution of leukocytes to pathology. Striking protective effects have been observed in a broad spectrum of inflammatory and immune reactions and ischemia-reperfusion syndromes (Table 3). These studies have provided convincing evidence that adhesion molecules are often critically involved in the pathogenesis of vascular and tissue damage. Moreover, they suggest that *anti-adhesion* therapy may be useful in the treatment of clinical disorders. Candidate therapeutics include *humanized* monoclonal antibodies, soluble receptors, peptide or saccharide antagonists that interfere with receptor-ligand interactions. Alternatively, it may be possible to develop small molecules that modulate the signal transduction pathways involved in the induction or activation of adhesion molecules.

CONCLUSION

In the past few years, there has been dramatic progress in the elucidation of the molecular basis of leukocyte adherence to endothelium; more than a dozen adhesion molecules involved in leukocyte-endothelial interactions have been immunologically characterized and molecularly cloned. These investigations have increased our understanding of the biology and clinical relevance of leukocyte adherence to endothelium in pathological conditions as well as normal physiology. Most importantly, a new approach to the treatment of a wide spectrum of clinical disorders may emerge from these studies.

REFERENCES

Anderson, D.C. and Springer, T.A., 1987, Leukocyte adhesion deficiency: An inherited defect in the Mac-1 and LFA-1 and p150, 95 glycoproteins. *Ann. Rev. Med.* 38:175-194.

Argenbright, L.W. and Barton, R.W., 1992, Interactions of leukocyte integrins with intercellular adhesion molecule 1 in the production of inflammatory vascular injury *in vivo*. *J. Clin. Invest.* 89: 259-272.

Atherton, A. and Born, G.V.R., 1973, Relationship between the velocity of rolling granulocytes and that of the blood flow in venules. *J. Physiol.* 157-165.

Bishop, M.J., Kowalski, T.F., Guidotti, S.M. and Harlan, J.M., 1992, Antibody against neutrophil adhesion improves reperfusion and limits alveolar infiltrate following unilateral pulmonary artery occlusion. *J. Surgical Res.* 52:199-204.

Butcher, E.C., 1991, Leukocyte-endothelial cell recognition: Three (or more) steps to specificity and diversity. *Cell* 67:1033-1036.

Carden, D.L., Smith, J.K. and Korthuis, R.J., 1990, Neutrophil-mediated microvascular dysfunction in postischemic canine skeletal muscle: Role of granulocyte adherence. *Circ. Res.* 66:1436-1444.

Carlos, T.M., and Harlan, J.M., 1990, Membrane proteins involved in phagocyte adherence to endothelium. *Immunologic Rev.* 114:5-28.

Clark, E.R. and Clark, E.L., 1935, Observations on changes in blood vascular endothelium in the living animal. *Am. J. Anat.* 57:385-438.

Clark, W.M., Madden, K.P., Rothlein, R. and Zivin, J.A., 1991, Reduction of central nervous system ischemic injury in rabbits using leukocyte adhesion antibody treatment. *Stroke* 22:877-883.

Clark, W.M., Madden, K.P., Rothlein, R. and Zivin, J.A., 1991, Reduction of central nervous system ischemic injury by monoclonal antibody to intercellular adhesion molecule. *J. Neurosurg.* 75-:623-627.

Cosimi, A.B., Conti, D., Delmonico, F.L., Preffer, F.I., Wee, S.L., Rothlein, R., Faanes, R. and Colvin, R.B., 1990, *In vivo* effects of monoclonal antibody to ICAM-1 (CD54) in nonhuman primates with renal allografts. *J. Immunol.* 144:4604-4612.

Diener, A.M., Beatty, P.G., Ochs, H.D. and Harlan, J.M., 1985, The role of neutrophil membrane glycoprotein 150(GP-150) in neutrophil-mediated endothelial cell injury *in vitro*. *J. Immunol.* 135:537-543.

Figdor, C.G. and van Kooyk, Y., 1992, Regulation of cell adhesion In: Harlan, J.M. and Liu, E. (eds); Adhesion: Its role in inflammatory disease. *W.H. Freeman & Co., New York*, 151-182.

Gundel, R.H., Wegner, C.D., Torcellini, C.A., Clarke, C.C., Haynes N., Rothlein, R., Smith, W. and Letts, L.G., 1991, Endothelial leukocyte adhesion molecule-1 mediates antigen-induced acute airway inflammation and late-phase airway obstruction in monkeys. *J. Clin. Invest.* 88:1407-1411.

Harlan, J.M., Winn, R.K., Vedder, N.B., Doerschuk, C.M. and Rice, C.L., 1992, *In vivo* models of leukocyte adherence to endothelium, In: Harlan, J.M. and Liu, E., (eds); Adhesion: Its role in inflammatory disease. *W.H. Freeman & Co., New York* 151-1826.

Harlan, J.M., Killen, P.D., Senecal, F.M., Schwartz, B.R., Yee, E.K., Taylor, R.F., Beatty, P.G., Price, T.A. and Ochs, H.D., 1985, The role of neutrophil membrane glycoprotein GP-150 in neutrophil adherence to endothelium *in vitro*. *Blood* 66:167-178.

Hermanowaski-Vosatka, A., Van Strijp, J.A.G., Swiggard, W.J. and Wright, S.D., 1992, Integrin modulating factor-1: A lipid that alters the function of leukocyte integrins. *Cell* 68:341-352.

Hernandez, L.A., Grisham, M.B., Twohig, G., Arfors, K.E., Harlan, J.M. and Granger, D.N. Role of neutrophils in ischemia-reperfusion induced microvascular injury. *Am. J. Physiol.* 253 (*Heart Circ. Physiol.* 22):H699-H703, 1987.

Horgan, M.J., Wright, S.D. and Malik, A.B., 1990, Antibody against leukocyte integrin (CD18) prevents reperfusion-induced lung vascular injury. *Am. J. Physiol.* 259:L315-L319.

Hutchings, P., Rosen, H., O'Reilly, L., Simpson, E., Siamon, G. and Cook, A., 1990, Transfer of diabetes in mice prevented by blockage of adhesion-promoting receptor on macrophages. *Nature* 346:636-642.

Issekutz, T.B. and Issekutz, A.C., 1990, T lymphocyte migration to arthritic joints and dermal- inflammation in the rat: Differing migration patterns and the involvement of VLA-4. *Clin. Immunol. Immunopathol.* 61:436-447.

Lawrence, M.B. and Springer, T.A., 1991, Leukocytes roll on a selectin at physiologic flow rates: distinction from and prerequisite for adhesion through integrins. *Cell* 65:859-873.

Ley, K., Gaehtgens, P., Fennie, C., Singer, M.S., Lasky, L.A. and Rosen, S.D., 1991, Lectin-like cell adhesion molecule 1 mediates leukocyte rolling in mesenteric venules *in vivo. Blood* 77:2553-2555.

Lorant, D.E., Patel, K.D., McIntyre, T.M., McEver, R.P., Prescott, S.M. and Simmerman, G.A., 1991, co-expression of GMP´-140 and PAF by endothelium stimulated by histamine or thrombin: A juxacrine system for adhesion and activation of neutrophils. *J. Cell Biol.* 115:223-234.

Ma, X.-L., Tsao, P.S. and Lefer, A.M., 1991, Antibody to CD18 exerts endothelial and cardiac protective effects in myocardial ischemia and reperfusion. *J. Clin. Invest.* 88:1237-1243.

Mileski, W.J., Winn, R.K., Vedder, N.B., Pohlman, T.H., Harlan, J.M. and Rice, C.L., 1990, Inhibition of CD18-dependent neutrophil adherence reduces organ injury after hemorrhagic shock in primates. *Surgery,* 108:206-212.

Movat, H.A., 1989, The role of endothelium in leukocyte emigration: The views of Cohnheim, Metchnikoff and their contemporaries. *Pathol. Immunopathol. Res.* 8:35-41.

Mulligan, M.S., Varani, J., Dame, M.K., Lane, C.L., Smith, C.W., Anderson, D.C. and Ward, P.A., 1991, Role of endothelial leukocyte adhesion molecule 1 (ELAM-1) in neutrophil-mediated lung injury in rats. *J. Clin. Invest.* 88:1396-1406.

Mulligan, M.S., Varani, J., Warren, J.S., Till, G.O., Smith, C.W., Anderson, D.C., Todd, R.F. III and Ward, P.A., 1992, Role of β_2 integrins of rat neutrophils in complement- and oxygen radical-mediated acute inflammatory injury. *J. Immunol.* 148:1847-1857.

Patarroyo, M., Lindbom, L. and Lundberg, C., 1991, Leukocyte adhesion: Molecular basis and relevance in inflammation. *Adv. Exp. Med. Biol.* 314:1-17.

Saez-Llorens, X., Jafari, H.S., Severien, C., Parras, F., Olsen, K.D., Hansen, E.J., Singer, I.I. and McCracken, G.H., Jr., 1991, Enhanced attenuation of meningeal inflammation and brain edema by concomitant administration of anti-CD18 monoclonal antibodies and dexamethasone in experimental *Haemophilus* meningitis. *J. Clin. Invest.* 88:2003-2011.

Simpson, P.J., Todd, R.F., III, Fantone, J.C., Mickelson, J.K., Griffin, J.D. and Lucchesi, B.R., 1990, Reduction of experimental canine myocardial reperfusion injury by a monoclonal antibody (anti-Mo1, anti-CD11b) that inhibits leukocyte adhesion. *J. Clin. Invest.* 81:624-629.

Smith, C.W., Marlin, S.D., Rothlein, R., Toman, C. and Anderson, D.C., 1989, Cooperative interactions of LFA-1 and Mac-1 with intercellular adhesion molecule-1 in facilitating adherence and transendothelial migration of human neutrophils *in vivo. J. Clin. Invest.* 83:2008-2107.

Springer, T.A., 1990, Adhesion receptors of the immune system. *Nature* 346:425-434.

Tuomanen, E.I., Saukkonen, K., Sande, S., Cioffe, C. and Wright, S.D., 1989, Reduction of inflammation, tissue damage, and mortality in bacterial meningitis in rabbit treated with monoclonal antibodies against adhesion-promoting receptors of leukocytes. *J. Exp. Med.* 170:959-968.

Vedder, N.B., Winn, R.K., Rice, C.L., Chi, E.Y., Arfors, K.E. and Harlan, J.M., 1990, Inhibition of leukocyte adherence by anti-CD18 monoclonal antibody attenuates reperfusion injury in the rabbit ear. *Proc. Natl. Acad. Sci. USA* 87:2643-2646.

Vedder, N.B., Winn, R.K., Rice, C.L., Chi, E. and Harlan, J.M., 1988, A monoclonal antibody to the adherence promoting leukocyte glycoprotein CD18 reduces organ injury and improves survival from hemorrhagic shock and resuscitation in rabbit. *J. Clin. Invest.* 81:939-944.

von Adrian, U.H., Chambers, J.D., McEvoy, L.M., Bargatze, R.F., Arfors, K.E. and Butcher, E.C., 1991, Two-step model of leukocyte-endothelial cell interaction in inflammation: Distinct roles for LECAM-1 and the leukocyte beta-2 integrins *in vivo. Proc. Natl. Acad. Sci. USA* 88:7538-754.

Wegner, C.D., Gundel, R.H., Reilly, P., Haynes, N., Letts, L.G. and Rothlein, R., 1990, Intercellular adhesion molecule-1 (ICAM-1) in the pathogenesis of asthma. *Science* 247:456-459.

Yednock, T.A., Cannon, C., Fritz, L.C., Sanchez-Madrid, F., Steinman, L. and Karin, N., 1992, Prevention of experimental autoimmune encephalomyelitis by antibodies against $\alpha_4\beta_1$ integrin. *Nature* 356:63-66.

Zimmerman, G.A., Prescott, S.M. and McIntyre, T.M., 1992, Endothelial interactions with granulocytes: tethering and signalling molecules. *Immunol. Today* 13:93-100.

MODULATION OF ENDOTHELIAL CELL FUNCTION BY CYTOKINES

M. Introna, F. Breviario, E. d´Aniello, E. Dejana
and Alberto Mantovani

Instituto di Ricerche Farmacologiche Mario Negri
Via Eritrea 62
20157 Milan
Italy

INTRODUCTION

The ontogeny and function of white blood cells require an intimate relationship with vascular endothelium. In the bone marrow, endothelial cells (EC), important producers of molecules with colony-stimulating (CSF) activity, are one crucial constituent of the hematopoietic milieu. In tissues, inflammatory and immunological reactions involve the active participation of EC, which interact closely with leukocytes.

The interplay between EC and leukocytes involves two modes of communication, one represented by physical interaction mediated by the regulated expression of receptor and counter-receptor molecules, and the other by the production of soluble mediators that reciprocally affect interacting cells.

This chapter will be mainly focused on how EC produce soluble polypeptide and respond to mediators, collectively referred to as cytokines.

Recent reviews of the interaction between vascular cells and leukocytes provide the background for the present chapter (*Mantovani and Dejana, 1989; Mantovani et al., 1992*). The functional programs activated in endothelial cells by different cytokines are discussed at length by *Mantovani et al., 1992*.

CYTOKINE PRODUCTION BY ENDOTHELIAL CELLS

IL-1, TNF and IL-6

IL-1, a cytokine that profoundly affects vascular cells (*see for review Mantovani and Dejana, 1989*) (Fig. 1), is also a product of vessel wall elements (*Warner and Libby, 1989*). Inducers of IL-1 production by vascular cells include LPS, TNF and IL-1 itself (*Warner and Libby, 1989; Maier et al., 1990*). *In vitro* passage of EC results in "spontaneous" expression of IL-1α message and refractiveness to activation by exogenous IL-1. This cytokine, which by itself is a weak proliferative signal for EC, inhibits EC proliferation induced by fibroblast growth factor (FGF) (*Cozzolino et al., 1990; Maier et al., 1990*). An antisense oligonucleotide complementary to IL-1α mRNA extends the *in vitro* life-span of EC (*Maier et al., 1990*). Collectively these results suggest that IL-1 produced by EC may represent an endogenous brake on EC proliferation and life-span.

Upon exposure to inflammatory signals and superinduction by protein synthesis inhibitors, vascular smooth muscle cells (SMC) can express TNF mRNA and protein

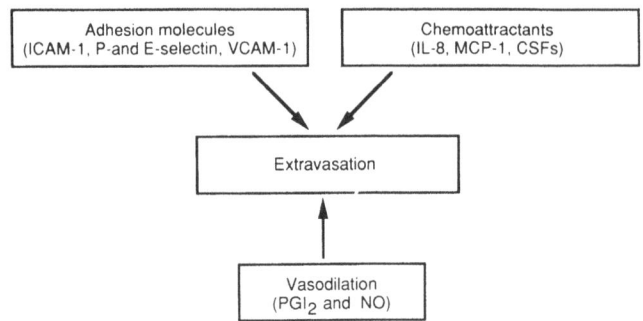

FIGURE 1: Mechanisms of TNF/IL-1 mediated leukocyte recruitment.

(*Warner and Libby, 1989*); however, we have consistently failed to detect TNF production in resting or LPS-stimulated EC (unpublished data).

The significance of TNF expression by vascular cells under extreme conditions remains to be defined.

SMC and EC produce high levels of IL-6 (*Sironi et al., 1989*). Variable levels of spontaneous IL-6 production can usually be detected, and these are markedly augmented by exposure to IL-1, TNF and LPS. IL-1 acts on T cells, at least in part, by inducing IL-6. Since IL-6 does not affect the pro-inflammatory and prothrombotic functions of IL-1 on EC, and anti-IL-6 antibodies do not affect the activity of IL-1, it can be concluded that IL-6 is not the ultimate mediator of the effects of IL-1 on vascular cells (*Sironi et al., 1989*).

IL-6 is a pleiotropic cytokine which affects T and B lymphocytes and the production of acute phase proteins by the liver. The production of IL-6 by EC integrates these cells in immunological circuits and in the regulation of acute phase response.

CSFs

EC are an important source of colony stimulating factors (CSFs). Hematopoietic growth factors produced by EC include granulocyte (G)-, granulocyte-macrophage (GM)- and monocyte (M)-CSF. CSF production is induced or augmented by a variety of stimuli, including LPS, IL-1, TNF (*for review see Mantovani and Dejana, 1989*) and minimally modified by low-density lipoproteins (*Rajavashist et al., 1990*). By producing CSFs, as well as IL-1 and IL-6 which have CSF activity, EC participate in the regulation of hematopoiesis. Since EC respond to certain CSF molecules (G- and GM-CSF and erythropoietin; see below), autocrine and paracrine circuits involving EC and CSFs could be important in the generation and maintenance of an appropriate bone marrow milieu.

Until recently, G- and GM-CSF were considered growth and differentiation factors restricted to hematopoietic elements. However, we found that EC express receptors for G- and GM-CSF and respond to these cytokines with migration and proliferation (*Bussolino et al., 1989*) and G-CSF was found to induce production of EC plasminogen activator (*Kojima et al., 1989*), an enzyme usually associated with cell migration. It was also observed recently that erythropoietin stimulates proliferation and chemotaxis of EC (*Anagnostou et al., 1990*). Unlike G- and GM-CSF, other molecules that share activity on hematopoietic precursors (M-CSF, IL-3, IL-6) had no effect on EC proliferation.

In an effort to define the pattern of activation of EC induced by CSF, a series of proliferation-independent functions were examined. Unlike IL-1 and interferon γ, used as reference cytokines in these studies, G- and GM-CSF did not affect expression of class II antigens of the major histocompatibility complex, or the adhesion proteins ICAM-1 and ELAM-1; nor did they induce pro-coagulant activity or PAF (*Bussolino et al., 1991*). These results are consistent with the hypothesis that G- and GM-CSF do not modulate EC functions related to inflammation/thrombosis or to accessory function. However, Chin *et al.* (1990) have shown that GM-CSF increases rat Peyer's patch high EC adhesiveness for lymphocytes.

The *in vivo* relevance of EC migration and proliferation induced by G- and GM-CSF was examined by evaluating angiogenesis in the rabbit cornea (*Bussolino et al., 1991*). G-CSF was used for these studies since this molecule is not species-restricted. It was found that G-CSF is an angiogenic stimulus, though less effective than FGF. Thus, G- and GM-CSF induce a set of functions in EC related to proliferation and migration and hence to the process of neovascularization. EC are an important component of the hematopoietic microenvironment, and the activity of the hematopoietic growth factors G- and GM-CSF on EC may contribute to the maintenance in the bone marrow of a microenvironment appropriate for hematopoiesis.

Chemotactic Cytokines

Vascular cells produce several members of a recently identified emerging superfamily of cytokines, many of which are chemotactic for leukocyte populations. The structural hallmark of these mediators is represented by four Cys residues, the first two in tandem, whose position is conserved and which probably determine a similar three-dimensional structure (*for review see Mantovani, 1990; Oppenheim et al., 1991*). EC produce IL-8 (also known as neutrophil-activating protein-1, NAP-1), and GRO, two members of the Cys-X-Cys family characterized by having a Cys tandem interrupted by an extra amino acid residue. IL-8 and GRO expression is induced in EC by inflammatory cytokines (IL-1 and TNF) and by LPS, and regulation is at the level of transcription (*Strieter et al., 1989; Sica et al., 1990a and 1990b; Standiford et al., 1990*). IL-8 and GRO are chemoattractants for neutrophils and, in addition to neutrophils, IL-8 affects lymphocytes, basophils (with IL-3) and melanoma cells (*Wang et al., 1990; Oppenheim et al., 1991*). As a result of proteolytic cleavage, IL-8 versions with different N-terminus and length can be produced, and it has been suggested that EC produce predominantly a 77 amino acid residue version of IL-8, which is substantially less active in activating leukocytes than the more common 73 amino acid residue molecule (*Gimbrone et al., 1989; Hebert et al., 1990*). Thrombin can catalyze the conversion to the smaller molecule (*Hebert et al., 1990*).

The effect of IL-8 on the interaction of neutrophils with EC has been the subject of apparently conflicting observations: on one hand, it was reported that IL-8 rendered neutrophils adhesive on normal EC (*Carveth et al., 1989*), whereas, EC-derived IL-8 was reported to inhibit binding of neutrophils to activated EC (*Gimbrone et al., 1989*). Further work is required to delineate the influence of experimental conditions (e.g. time of interaction), specificity (e.g. effect of classical chemoattractants such as FMLP and C5a) and *in vivo* relevance of these observations.

Upon exposure to inflammatory signals (IL-1, TNF and LPS), EC produce monocyte chemotactic protein (MCP) (*Sica et al., 1990b*), a member of the Cys-X-Cys family. As for IL-8, regulation of MCP in EC is at the level of gene transcription. MCP is a chemoattractant active on monocytes, *in vitro* and *in vivo*, and it is important in the regulation of monocyte infiltration in certain tumors (*Mantovani, 1990*). In addition to EC, SMC and ontogenetically related renal mesangial cells also produce MCP, and production is induced also by minimally modified low-density lipoproteins (*Cushing et al., 1990; Zoja et al., 1991; Wang et al., 1991*).

Thus, vascular cells produce chemoattractants (e.g. IL-8 and MCP), which attract and activate different leukocyte populations. These mediators act in concert with the *de novo* or augmented expression of adhesion molecules at sites of inflammation plus vasodilators (see above) to cause extravasation in tissues of circulating cells (Fig. 1). Metastatic tumor cells are affected, as well as leukocytes.

The production of chemoattractants by vascular cells could also be important in the pathogenesis of diseases that affect the vessel wall. For example, monocyte infiltration is an early event in the natural history of atherosclerosis and a prominent feature of vasculitis, and it is likely that local production of MCP, elicited by modified low-density lipoproteins or inflammatory signals, plays a role in these diseases.

TABLE 1: Clones isolated by differential screening

Gene	Number of positive clones	Length of insert (base pairs)	Function	Reference
IL-8	16	800	Chemoattractant	Strieter et al., 1989
ELAM-1	07	1500	Adhesion Molecule	Bevilacqua et al., 1989
GRO-α	06	800	Chemoattractant Cytokine	Wen et al., 1989
GRO-β	05	900	Chemoattractant cytokine	Haskill et al., 1990
PA-1	02	1400	Inhibitor of plasminogen	Gramse et al., 1986
I-KB	02	600	Inhibitor of transcription	Haskill et al., 1991
PTX3	02	900	New member of the Pentaxin gene	Breviario et al., 1992

SEARCHING FOR NOVEL IL-1-INDUCIBLE GENES

As discussed in detail elsewhere (*Mantovani et al., 1992*), IL-1 and the functionally related cytokine TNF activate EC functions mainly related to inflammation and thrombosis.

In order to systematically explore the genetic background of this response, we have decided to use a molecular biology approach, that is to clone the genes which are transcriptionally induced in the endothelial cells upon exposure to IL-1. We started to look at the genes that are induced rapidly (after 1 hour exposure to IL-1) and in the absence of protein synthesis, thus concentrating on the so called "immediate-early response genes".

A cDNA library was constructed from poly (A+) mRNA isolated from human umbilical vein endothelial cells (HUVEC) stimulated with 20ng/ml human recombinant IL-Iβ for one hour in the presence of cycloheximide. Over 4000 clones were individually screened by differential hybridization with cDNA probes derived from untreated or IL-Iβ treated HUVEC (*Breviario et al., 1992*). Forty (40) clones were identified (1%), which gave a stronger signal with the cDNA from IL-1β-treated cells. Partial sequencing of these clones revealed that 38 of them corresponded to genes already cloned, that is Interleukin-8 (IL-8) (16 clones) (*Strieter et al., 1989*), endothelial leukocyte adhesion molecule 1 (ELAM-1) (*Bevilacqua et al., 1989*) (7 clones), Gro-a (6 clones) (*Wen et al., 1989*), Gro-D (5 clones) (*Haskill et al., 1990*), plasminogen activator inhibitor (PA-i) (2 clones) (*Gramse et al., 1986*) and the inhibitor of the transcription factor NF-kB (I-kB) (2 clones) (*Haskill et al., 1991*) (Table 1). It is interesting to observe that with the exception of the I-kB gene all other cloned genes had been previously reported to be induced in HUVEC by IL-1 (6-10), and in fact may belong to the prothrombotic, proinflammatory program (Table 1). The induction of I-kB, a negative regulator of a transcriptional factor may provide a genetic background for the observed inhibition of several functions in HUVEC upon exposure to IL-1.

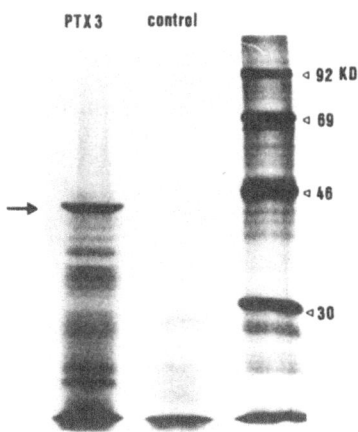

PTX3   control

◁ 92 KD

◁ 69

◁ 46

◁ 30

FIGURE 2: *In vitro* transcription and translation of PTX3. The PTX3 cDNA insert was subcloned in the pGem3 vector downstream from the SP6 promoter. After linearization of the plasmid with KpnI, *in vitro*, transcription was carried out *in vitro* with the SP6 RNA polymerase. The *in vitro* made RNA served as template for *in vitro* translation in presence of wheat germ extract and ^{35}S- labelled methionine. As control, the transcription and translation reactions were carried with the empty pGem3 plasmid. The products were run in a 10% SDS/polyacrylamide gel and autoradiographed.

One such candidate is the ELAM-1 gene itself, which is rapidly induced and then repressed (*Bevilacqua et al.*, 1989) and contains a functional NF-kB binding site in its promoter region (*Whelan et al., 1991*). The last two clones contained an insert of 900 base pairs belonging to a previously unknown gene, as determined by screening of the EMBL Nucleotide Sequence Database. Since it contained an incomplete open reading frame, it was necessary to screen other HUVEC cDNA libraries. The full-length cDNA was isolated and called PTX3. It contains a 68 base pair 5′ untranslated region, an open reading frame of 1143 base pairs, coding for 381 amino acids, and a TAA stop codon, followed by a polyadenylation signal (data not shown).

The 3′ terminal half of the predicted protein can be significantly aligned with the nine cloned members of the pentaxin gene family, including the two human members C-reactive protein (CRP) and serum amyloid P component (SAP) (*Pepys and Baltz, 1983*). There is an overall identity of 22 amino acids over 259 consensus length (9%) and a similarity of 55 amino acids (21%) (data not shown).

In particular, PTX3 contains 8 amino acids (H-L-C-G-T-W-N-S), which fit perfectly the pentaxin family signature (H-X-C-X-S/TW-X-s) as identified by the PROSITE protein pattern database (*Bairoch, 1991*). Finally, two cysteines at position 210 and 271 of PTX3 are conserved in all members of the family, from horseshoe crab to man (data not shown).

The 5′ end portion of the protein contains a 17 amino acids signal peptide, indicating that the protein product may be secreted. No other motif or homology could be identified in this region, which has a very hydrophilic profile (data not shown).

In order to verify that the full length cDNA of PTX3 contains all the necessary elements of a functional gene, we have subcloned it in a pGem3 vector. The cDNA was transcribed and translated *in vitro* with the SP6 RNA polymerase and wheat germ extract in the presence of 35S-methionine. The products were separated in a SDS-polyacrylamide gel and the autoradiograph is shown in Figure 2. The most abundant product was a 42 kD protein, which was absent from the negative control lane in complete agreement with the predicted molecular weight of a protein starting from the most 5´ methionine of the full length cDNA clone.

These preliminary observations show that we have cloned a full length cDNA from IL-1β-stimulated HUVEC for a novel gene, which we called PTX3. In the 3´ half of the predicted protein it shows a high degree of homology with all the known members of the pentaxin gene family including CRP and SAP. It is, therefore, tempting to speculate that PTX3 may also represent, like CRP and SAP, an acute phase response gene which would be in agreement with its inducibility by IL-1. Furthermore, its presence in endothelial cells may complement the classic pentaxins and represent a novel indicator of tissue reactions, particularly of those involving the vessel wall.

CONCLUDING REMARKS

EC and leukocytes interact closely in the activation and expression of immunity as well as in inflammatory reactions and hemostasis, and cytokines are one important communication system between leukocytes and EC (*Mantovani, 1992*). A second mode of communication is represented by the regulated expression of receptors and their ligands, the surface of vessel wall cells and white blood cells, which permits the physical interaction of these cell types. Vascular cells are both a target for and a source of soluble polypeptide and lipid mediators.

The identification of novel genes expressed in activated EC, such as PTX3, may provide novel insights and diagnostic markers for vascular diseases.

REFERENCES

Anagnostou, A., Lee, E.S., Kessimian, N., Levinson, R. and Steiner, M., 1990, Erythropoietin has a mitoogenic and positive chemotactic effect on endothelial cells. *Proc. Natl. Acad. Sci. U.S.A.* 87:5978.

Bairoch, A., 1991, PROSITE:A dictionary of sites and patterns in protein. *Nucleic Acids Res.* 19 (suppl.):2241.

Bevilacqua, M.P., Stengelin, S., Gimbrone, M.A. Jr. and Seed, B., 1989, Endothelial leukocyte adhesion molecule 1: An inducible receptor for neutrophils related to complement regulatory proteins and lectins. *Science* 243:1160.

Breviario, F., D´Aniello, E. Golay, J., Peri, G., Bottazzi, B., Bairoch, A., Saccone, S., Marzella, R., Predazzi, V., Gocchi, M., Della Valle, G., Dejana, E., Mantovani, A. and Introna, M., 1992, Interleukin-1 inducible genes in endothelial cells: Cloning of a new gene related to Creative protein and serum amyloid P component. *Submitted for Publication*.

Bussolino, F., Wang, J.M., Defilippi, P., Turrini, F., Sanavio, R., Edgell, C.J., Aglietta, M., Arese, P. and Mantovani, A., 1989, Granulocyte- and granulocyte-macrophage colony stimulating factors induce human endothelial cells to migrate and proliferate. *Nature* 337:471.

Bussolino, F., Ziche, M., Wang, J.M., Alessi, D., Morbidelli, L., Cremona, O., Bosia, A., Marchisio, P. C. and Mantovani, A., 1991, *In vitro* and *in vivo* activation of endothelial cells by colony-stimulating factors. *J. Clin. Invest.* 87:986.

Carveth, H.J., Bohnsack, J.F., McIntyre, T.M., Baggioline, M. Prescott, S.M. and Zimmerman, G.A., 1989, Neutrophil activating factor (NAF) induces polymorphocyclear leukocyte adherence to endothelial cells and to subendothelial matrix proteins. *Biochem. Biophys. Res. Commun.* 162:387.

Chin, Y-H, Cai, J-P and Johnson, K., 1990, Lymphocyte adhesion to cultured Peyer's patch high endothelial venule cells is mediated by organ-specific homing receptors and can be regulated by cytokines. *J. Immunol.* 145:3669.

Cozzolino, F., Torcia, M., Aldinucci, D., Ziche, M., Almerigigna, F., Bani, D. Stern, D.M., 1990, Interleukin-1 is an autocrine regulator of human endothelial cell growth. *Proc. Natl. Acad. Sci. U.S.A.* 87:6487.

Cushing, S.D., Berliner, J.A., Valente, A.J., Territo, M.D., Navab, M., Parhami, F., Gerrity, R., Schwartz, C.J. and Rogelman, A.L.M., 1990, Minimally modified low density lipoprotein induces monocyte chemotactic protein 1 in human endothelial cells and smooth muscle cells. *Proc. Natl. Acad. Sci. U.S.A.* 87:5134.

Gimbrone, M.A. Jr., Obin, M.S., Brock, A.F., Luis, E.A., Hass, P.E., Hebert, C.A., Yip, Y.K., Leung, D.W., Lowe, D.G., Kohr, W.J., Darbonne, W.C., Bechtol, K.B., and Baker, J.B., 1989, Endothelial interleukin 8:An novel inhibitor of leukocyte-endothelial interactions. *Science*, 246:1601.

Granse, M., Breviario, F., Pintucci, G., Millet, I., Dejana, E., Van Damme, J., Donati, M.B. and Muss one, L., 1986, Enhancement by interleukin-1 (IL-1) of plasminogen activator inhibitor (PA-I) activity in cultured human endothelial cells. *Biochem. Biophys. Res. Commun.* 139:720.

Haskill, S., Beg, A.A., Tompkins, S.M., Morris, J.S., Yurochko, A.D., Sampson-Johannes, A., Mondal, K., Ralph, P. And Baldwin, A.S. Jr., 1991, Characterization of an immediate-early gene induced in adherent monocytes that encodes IkB-like activity. *Cell* 65:1281.

Haskill, S., Peace, A., Morris, J., Sporn, S.A., Anisowicz, A., Lee, S.W., Smith, T., Martin, G., Ralph, P. and Sager, R., 1990, Identification of three related human gro genes encoding cytokine functions. *Proc. Natl. Acad. Sci. U.S.A.* 87:7732.

Hébert, A., Luscinskas, F.W., Kiely, J.M., Luis, A., Darbonne, C., Bennet, L., Liu, C., Obin, S., Gim brone, A. Jr. and Baker, B., 1990, Endothelial and leukocyte forms of interleukin 8:Conversion by thrombin and interactions with neutrophils. *J. Immunol.* 145:3033.

Kojima, S., Tadaenuma, H., Inanda, Y. and Saito, Y., 1989, Enhancement of plasminogen activator activity in cultured endothelial cells by granulocyte colony-stimulating factor. *J. Cell Physiol.* 138:192.

Maier, J.A.M., Voulalas, P., Roeder, D. and Maciag, T., 1990, Extension of the life-span of human endothelial cells by an interleukin-1α antisense oligomer. *Science* 249:1570.

Mantovani, A., 1990, Tumor-associated macrophages. *Curr. Opin. Immunol.* 2:689.

Mantovani, A., Bussolino, F. and Dejana, E., 1992, Cytokine regulation of endothelial cell function. *FASEB J.* 6:2591.

Mantovani, A. and Dejana, E., 1989, Cytokines as communication signals between leukocytes and endothelial cells. *Immunol. Today* 10:370.

Oppenheim, J.J. Zachariae, C.O.C., Mukaida, N. and Matsuchima, K., 1991, Properties of the novel proinflammatory supergene "Intercrine: cytokine family. *Annu. Rev. Immunol.* 9:617.

Pepys, M.B. and Baltz, M.L., 1983, Acute phase proteins with special reference to C-reactive protein and related proteins (pentaxins) and serum amyloid A protein. *Adv. Immunol.* 34:141.

Rajavashisth, T.B., Andalibi, A., Territo, M.C., Berliner, J.A., Navah, M. Fogelman, A.M. and Lusis, A.J., 1990, Induction of endothelial cell expression of granulocyte and macrophage colony-stimulating factors by modified low-density lipoproteins. *Nature* 344:254.

Sica, A., Matsushima, K., Van Damme, J., Wang, J.M., Polentarutti, N., Dejana, E., Colotta, F. and Mantovani, A., 1990a, IL-1 transcriptionally activates the neutrophil chemotactic factor/IL-8 gene in endothelial cells. *Immunology* 69:548.

Sica, A., Wang, J.M., Colotta, F., Dejana, E., Mantovani, A., Oppenheim, J.J., Larsen, C.G., Zacha riae, C.O.C. and Matsushima, K., 1990b, Monocyte chemotactic and activating factor gene expression induced in endothelial cells by IL-1 and TNF. *J. Immunol.* 144:3034.

Sironi, M., Breviario, F., Proserpio, P., Biondi, A., Vecchi, A., Van Damme, J., Dejana, E. and Mantovani, A., 1989, IL-1 stimulates IL-6 production in endothelial cells. *J. Immunol.* 142:549.

Standiford, T.J., Strieter, R.M., Kasahara, K. and Kunkel, S.L., 1990, Disparate regulation of interleukin 8 gene expression from blood monocytes, endothelial cells, and fibroblasts by interleukin 4. *Biochem. Biophys. Res. Commun.* 171:531.

Strieter, R.M., Kunkel, S.L., Showell, J.H.J. Remick, D.G., Phan, S.H., Ward, P.A. and Marks, R.M., 1989, Endothelial cell gene expression of neutrophil chemotactic factor by TNF-α, LPS and IL-1β. *Science* 243:1467.

Wang, J.M., Sica, A., Peri, G., Walter, S., Martin Padura, I., Libby, P., Ceska, M., Lindley, I., Colotta, F. and Mantovani, A., 1991, Expression of monocyte chemotactic protein and interleukin-8 by cytokine-activated human vascular smooth muscle cells. *Arteriosclerosis* 11:1166.

Wang, J.M., Taraboletti, G., Matsuchima, K., Van Damme, J. and Mantovani, A., 1990, Induction of haptotactic migration of melanoma cells by neutrophil activating protein/interleukin-8. *Biochem. Biophys. Res. Commun.* 169:165.

Warner, S.J.C. and Libby, P., 1989, Human vascular smooth muscle cells: Target for and source of tumor necrosis factor. *J. Immunol.* 142:100.

Wen, D., Rowland, A. and Derynck, R., 1989, Expression and secretion of gro/MGSA by stimulated human endothelial cells. *EMBO J.* 8:1761.

Whelan, J., Ghersa, P., van Huijsduijnen, R.H., Gray, J., Chandra, G., Talabot, F. and DeLamarter, J.F., 1991, An NFkB-like factor is essential but not sufficient for cytokine induction of endothelial leukocyte adhesion molecule 1 (ELAM-1) gene transcription. *Nucleic Acids Res.* 19:2645.

Zoja, C., Wang, J.M., Bettoni, S., Sironi, M., Renzi, D., Chiaffarino, F., Abboud, H.E., Mantovani, A., Remuzzi, G and Rambaldi, A., 1991, Interleukin-1β and tumor necrosis factor-α induce gene expression and production of leukocyte chemotactic factors, colony stimulating factors, and interleukin-6 in human mesangial cells. *Am. J. Pathol.* 138:991.

CIRCULATING ENDOTHELIAL ADHESION MOLECULES

John L. Gordon, R.M. Edwards, S.J. Cashman*,
A.J. Rees* & A.J.H. Gearing

British Bio-technology Limited
Watlington Road
Cowley, Oxford OX4 5LY
United Kingdom

*Royal Postgraduate Medical School
Hammersmith Hospital, London W12 ONN
United Kingdom

INTRODUCTION

Activation and damage of the vascular endothelium is a feature of inflammatory conditions such as atherosclerosis, diabetes, septic shock, acute respiratory distress syndrome and graft rejection. The ability to monitor the state of the endothelium *in vivo* would be useful in the management of such diseases. Recently, a number of activation antigens which mediate leucocyte or tumor cell adhesion have been defined on human umbilical vein endothelial cells in culture. These activation antigens have been cloned and sequenced and include Intercellular adhesion molecule-1 (ICAM-1) (*Simmons et al., 1988*) and Vascular cell adhesion molecule-1 (VCAM-1)(*Osborn et al., 1989*) which are members of the immunoglobulin superfamily, and E-selectin (also known as ELAM-1) (*Bevilacqua et al., 1989*) and P-selectin (also known as GMP-140)(*Johnston et al., 1989*), which belong to a new family of adhesion molecules called the selectins. The selectins are characterized by an N-terminal c-type lectin domain closely linked to an epidermal growth factor domain, several consensus repeat domains similar to those found in complement regulatory proteins, a transmembrane domain and intracytoplasmic tail. The properties of these adhesion molecules are listed in Table 1. Current models implicate the selectins in the initial capture and rolling reactions that leukocytes are seen to undergo on vascular endothelium, both *in vitro* and *in vivo*, whereas interactions between the leucocyte integrins and immunoglobulin superfamily adhesion molecules are involved in subsequent migration and extravasation (*Lawrence and Springer, 1991*). *In vivo* studies have shown that blocking the function of these adhesion molecules can prevent leucocyte extravasation and inflammation thereby confirming their significance in the inflammatory process (*Von Andrian et al., 1991; Watson et al., 1991; Wegner et al., 1990 ; Mulligan et al., 1991*).

Expression of adhesion molecules on the endothelium has been demonstrated by immunohistology in a number of diseases (*Adams et al., 1989; Faull and Russ, 1989, Leung et al., 1991; Norris et al., 1991; Rice et al., 1991; Ruco et al 1991; Ruco et al., 1992*). However, histological studies provide little information about the dynamics of disease processes, and routine monitoring of patients by biopsy is not normally possible. Consequently, serum markers of endothelial damage have been sought. Soluble forms of endothelial surface proteins have been described in serum, but so far only those which are also found on circulating cells, e.g. P-selectin (which is released

Vascular Endothelium, Edited by J.D. Catravas *et al.*
Plenum Press, New York, 1993

TABLE 1: Properties of the vascular endothelial adhesin molecules, ICAM-1, VCAM-1, E-selectin and P-selectin

Property	ICAM-1	VCAM-1	E-Selectin	P-Selectin
Structure	5 Ig domains	7 & 6 Ig domains	Selectin 6 CRP	Selectin 9/10 CRP
Cell surface	Broad 96 kD	Endothelium Dendritic, neuronal 106 kD	Endothelium 117 kD	Endothelium Platelets 140 kD
Soluble form	89 kD	100 kD	110 kD	137 kD
Ligands	LFA-1,MAC-1,CD43	VLA-4	sLex,sLea	sLex,sLea

by platelets) and ICAM-1 (which is also released by leukocytes). We therefore developed specific two-site ELISAs for the detection of soluble forms of E-selectin (endothelial specific) and VCAM-1 (restricted to endothelium and some tissue macrophages and neuronal cells). We also developed two-site ELISAs for ICAM-1 to extend previous studies and compare ICAM-1 data with those for E-selectin and VCAM-1. In particular, we investigated the release of adhesion molecules from endothelial cells and melanoma cells *in vitro* and have monitored their levels in blood of normal individuals and patients with immunological and inflammatory diseases.

METHODOLOGY

Concentrations of soluble E-Selectin, VCAM-1 and ICAM-1 were measured using dual monoclonal antibody two-site ELISAs. The antibodies used in the assays were generated as described in Pigott *et al.*, 1991 and were specific for E-Selectin, VCAM-1 or ICAM-1. Assays were calibrated using recombinant soluble E-Selectin or VCAM-1 lacking their transmembrane and cytoplasmic domains and a detergent solubilized extract of CHO cells expressing a recombinant ICAM-1 construct. Standards were aliquoted and stored at -20°C and given an arbitrary unitage against which all samples were measured.

Specificity of the assays was confirmed by titration of each of the three recombinant standards. The E-selectin assay produces a dose response from 0.2 to 3.0 units/ml with recombinant E-selectin and no response to 100μg/ml P-selectin (Takara) 100 units/ml of ICAM-1 or VCAM-1. One unit of E-selectin corresponds to approximately 2 ng of purified recombinant soluble E-selectin by ELISA. The VCAM-1 assay gave a dose response curve from 1.5 to 50 units/ml and no reaction to E-selectin or ICAM-1. One unit of VCAM-1 corresponds to approximately 9 ng of purified recombinant VCAM-1 by ELISA. The ICAM-1 assay gave a dose response curve from 2.0 to 30 units/ml and no reaction to E-selectin or VCAM-1 by ELISA. One unit of ICAM-1 corresponds to 4.5 ng of purified recombinant solubilized ICAM-1.

EVIDENCE FOR RELEASE OF SOLUBLE ADHESION MOLECULES *IN VITRO*

Leukocytes

Leukocytes have been shown to release several adhesion molecules from their

TABLE 2: Levels of soluble adhesins in human serum

SAMPLE	ICAM-1	VCAM-1	E-SELECTIN
Normal n=62	56.4 ± 56.4	54.7 ± 21.5	10.8 ± 5.9
Diabetes n=55	110.4 ± 67.8	116.0 ± 82.6	23.2 ± 13.0
IRF n=24	108.9 ± 67.5	112.0 ± 41.9	16.7 ± 10.8
CAPD n=9	141.2 ± 81.0	118.3 ± 46.3	13.1 ± 11.3
Hypertension n=10	69.5 ± 50.2	73.9 ± 35.8	15.2 ± 8.8

Values quoted are in mean Units/ml ± standard deviation

cell surfaces following activation. Thus, L-selectin is rapidly shed by neutrophils following activation with cytokines or PAF (*Kishimoto et al., 1989; Von Andrian et al., 1991*). The mechanism of release is not known, though neutrophils are known to release soluble CD43 (a ligand for ICAM-1) and CD44, a leucocyte homing receptor, by a protease dependent process (*Campanero et al., 1991*). Soluble ICAM-1 has been reported to be produced by peripheral blood mononuclear cells and by B-lymphoblastoid cell lines, but not by endothelial cells, adenocarcinoma, fibroblasts, T lymphoma or myelomonocytic cell lines (*Rothlein et al., 1991*).

Platelets

Platelets can release a soluble form of P-selectin following stimulation with thrombin *in vitro* (*Dunlop et al., 1992*). An alternatively spliced soluble variant of P-selectin has been described (*Johnston et al., 1989*). Messenger RNA for soluble and cell surface forms of P-selectin are found in equal amounts in platelets (*Johnston et al., 1990*)

Tumor cells

We have demonstrated previously that human melanoma cells can be stimulated by cytokines to express ICAM-1 on their cell surface and to release soluble ICAM-1 into the supernatant. In contrast, colon carcinoma cells can be stimulated to express equivalent levels of cell-surface ICAM-1 but without concomitant production of soluble ICAM-1 (*Giavazzi et al., 1992*). The ICAM shed by melanoma cells has been shown to block natural killer and lymphokine activated killer cell cytotoxicity (*Becker et al., 1991*).

Several human carcinoma cell lines have also been shown to release soluble ICAM-1 in culture following activation with interferon gamma (*Tsujisaki et al., 1991*)

Endothelial cells

We have investigated the release of soluble ICAM-1, VCAM-1 and E-selectin by unstimulated and cytokine-activated HUVECs in culture. Unstimulated cells do not release detectable levels of any of these adhesins into the culture supernatant. Following activation with IL-1 and TNF, however, we can detect ICAM-1, VCAM-1 and E-selectin in the cell-free supernatants (*Pigot et al., 1992*). The mechanism of release is at present unclear, though immunoprecipitation analysis reveals that the soluble forms are of lower Mr than the cell surface molecules (Table 1), consistent with proteolytic cleavage close to the transmembrane domains. In contrast to P-selectin, there are no published reports of alternatively spliced soluble variants of ICAM-1, VCAM-1 or E-selectin.

EVIDENCE FOR *IN VIVO* RELEASE

Animal Models

We have shown that human melanoma cells grown as tumors in nude mice can release soluble ICAM-1 into the blood (*Giavazzi et al., 1992*). The level of ICAM-1 in the serum showed a positive correlation with the weight of the tumor.

Normal Individuals

Normal, apparently healthy individuals have been shown to have detectable levels of ICAM-1 (mean 156 ng/ml, n = 10 (*Rothlein et al., 1991*) and P-selectin (mean 251 ng/ml in males and 175 ng/ml in females, n = 10 (*Dunlop et al., 1992*)) in their serum.

In order to confirm and extend these studies, we studied a large control group consisting of 26 blood donors and 36 healthy laboratory personnel. Mean concentrations for E-selectin were 10.8 ± 5.9 units/ml (mean ± SD, range 0-33; 95% confidence limit <22.6); for VCAM-1 were 54.7 ± 21.5 units /ml (range 25-120 , 95% confidence limit < 97.7); and for ICAM-1 were 56.4 ± 31 units/ml (range 13-160, 95% confidence limit <118.5) (see Table 2). These mean values correspond to 21.6 ng/ml of E-selectin, 492 ng/ml VCAM-1 and 254 ng/ml ICAM-1.

The use of two-site ELISAs shows that the soluble adhesion molecules are immunologically intact but does not necessarily confirm their identity. P-selectin has been purified from serum, shown to have an authentic N-terminus by microsequencing, and demonstrated to mediate neutrophil adhesion (*Dunlop et al., 1992*). The purified material has a similar Mr to cell membrane P-selectin but migrates as a monomer as opposed to the tetrameric membrane form. ICAM-1 can also be purified from serum and shown to mediate leucocyte adhesion via the ICAM-1 ligand LFA-1 (*Rothlein et al., 1991*). Soluble ICAM migrates at 80 kDa under reducing and denaturing conditions and at greater than 200 kDa by gradient gel electrophoresis (*Seth et al., 1991*). We have purified VCAM-1 from serum and confirmed its identity by amino acid sequencing.

The presence of soluble adhesion molecules in the blood of apparently healthy individuals is perhaps unexpected, given that their expression *in vitro* is dependent on activation with inflammatory stimuli. It is possible that release occurs from sites, such as lung, gut or hematopoietic tissues, as a result of multiple minor inflammatory events that occur during normal activity. Thus, the cell source for each of these adhesion molecules may differ. ICAM-1 can be derived from many cell types, including leukocytes, connective tissues and endothelium (*Springer, 1990*), VCAM-1 can be produced by dendritic, monocytic and neuronal cells (*Birdsall et al., 1992*), whereas E-selectin is only produced by endothelium (*Bevilacqua et al., 1989*).

CLINICAL STUDIES

Patients who would be predicted to have elevated levels of soluble endothelial adhesion molecules include those with widespread microvascular injury, such as in diabetes, or those in whom the endothelium is exposed to pro-inflammatory cytokines, such as in graft rejection, chronic inflammation or septic shock. We have shown previously that patients undergoing repeated hemodialysis have high E-selectin, VCAM-1, and ICAM-1 concentrations in their serum (*Gearing et al., 1992*), possibly because of *ex vivo* exposure of blood to dialysis membranes which has been shown to provoke TNF and IL-1 release from leukocytes (*Ryan et al., 1991*).

Other groups have shown that serum levels of soluble ICAM-1 can be elevated above those seen in normal individuals in melanoma (*Harning et al., 1991*), metastatic carcinoma (*Tsujisaki et al., 1991*), leucocyte adhesion deficiency (*Rothlein et al., 1991*), renal colic (*Seth et al., 1991*) and cardiac allograft rejection (*Ballantyne et al., 1991*). We have studied sera from a number of patient groups including hypertensives, diabetics, and patients with impaired renal function (IRF) or on chronic ambulatory peritoneal dialysis (CAPD) (see Table 2.). Levels of circulating E-selectin were preferentially elevated in diabetic patients and in IRF patients. VCAM-1 was elevated

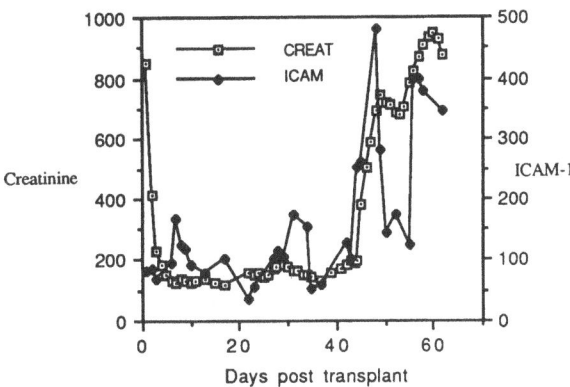

Figure 1. Renal Transplants: Serum Creatinine
and Soluble ICAM-1 levels.

in patients with IRF, CAPD and in diabetics. ICAM-1 was elevated in all of the groups except for patients with hypertension who had levels of all soluble adhesion molecules within the normal range.

Soluble adhesin concentrations did not correlate with those of glucose or fructosamine as indices of short or medium term diabetic control, nor were they correlated with serum creatinine or CRP in either diabetics or non diabetics. Thus, the elevated levels were not simply a reflection of impaired renal function, glycemic control or an acute phase response. Other serum markers of endothelial damage, including Factor VIII related antigen and endothelin, have also been shown to be elevated in diabetes and renal disease (*Brown et al., 1986; Pasi et al., 1990 ; Warrens et al., 1990*). The serum of diabetic rats contains an activity of greater than 12 kDa, which reduces the ability of leukocytes to roll on the vascular endothelium (*Fortes et al., 1991*). Given our demonstration of elevated E-selectin in diabetic patient sera and our demonstration that soluble E-selectin can inhibit leucocyte adhesion (see below), it is possible that the activity described by Fortes *et al.* could be due to soluble E-selectin in the serum.

In order to determine the prognostic significance of measuring levels of soluble adhesion molecules, we have begun longitudinal studies in individual renal allograft patients. The clinical history of these patients is well documented, and regular serum samples are taken to monitor creatinine levels. Initial results indicate that rejection episodes, as defined by a rise in serum creatinine and confirmed by biopsy, are accompanied by a rise in levels of ICAM-1, VCAM-1 and E-selectin in the serum. Figure 1 illustrates the results for ICAM-1 in one such patient. In some cases, the rise in levels of soluble adhesion molecules precedes the rise in creatinine levels. In some patients, a rise in serum VCAM was associated with a cytomegalovirus infection. Although these results are very encouraging, further studies are required before a prognostic or diagnostic role for measurement of soluble adhesins can be confirmed.

BIOLOGICAL ROLE OF SOLUBLE ADHESION MOLECULES

The biological significance of circulating soluble adhesins is unknown, but there are a number of possible functions:

1. **Down-modulation of adhesion:** Soluble adhesion molecules in the blood may prevent adhesion of leukocytes to endothelium by binding to their cell-surface ligands; we have observed this *in vitro* in experiments with recombinant soluble E-selectin and VCAM-1 (see methodology). Removal of adhesion molecules from the endothelial

surface would also serve to reduce the adhesiveness of the endothelium. This may be a general mechanism operating in all inflammatory reactions essential to allow the release of leukocytes from the endothelium into the underlying tissues.

The observation that melanoma cells secrete ICAM-1, which can block cell killing by cytotoxic lymphocytes, suggests that this could also be an escape mechanism used by tumors to avoid immune surveillance.

2. Cytokine function: Binding of soluble forms of adhesion molecules to their ligands on the leucocyte could transmit a signal and affect cell function. For example, a recombinant form of E-Selectin has been shown to be a neutrophil chemoattractant (*Lo et al., 1991*) and soluble forms of VCAM-1 and ICAM-1, when coated on plastic, can co-stimulate lymphocytes (*Van Seventer et al., 1992*).

3. Anti-viral function: Recombinant soluble ICAM-1 can block rhinovirus adhesion, suggesting that natural soluble ICAM-1 may have anti-viral activity (*Marlin et al., 1990*).

4. Bystander effect: Soluble adhesion molecules may be released simply as a consequence of protease production by leukocytes during the process of extravasation and serve no other function.

The precise role of soluble adhesins will only be determined when the mechanisms which control their release are defined and can be manipulated. Patients with elevated levels of E-selectin do not necessarily have elevated levels of VCAM-1 or ICAM-1 and vice versa. This suggests differential control of expression and release *in vivo*. If release of adhesion molecules is shown to be a necessary component of leucocyte extravasation, then its modulation could represent a useful target for therapeutic intervention.

REFERENCES

Adams, D.H., Hubscher, S.G., Shaw, J., Rothlein, R. & Neuberger, J.M., 1989, ICAM-1 on liver allografts during rejection. *Lancet* 2: 1122-1125.

Ballantyne, C., Mainolfi, E.A., Young, J.B., Windsor, N.T., Cocanougher, B., Lawrence, E.C., Anderson, D.C. & Rothlein, R., 1991, Prognostic value of increased levels of circulating ICAM-1 after heart transplant. *Clin. Res.* 39:285a.

Bevilacqua, M.P., Stengelin, S., Gimbrone, M.A., Seed, B., 1989, Endothelial leucocyte adhesion molecule 1: An inducible receptor for neutrophils related to complement regulating proteins and lectins. *Science* 243:1160-61.

Becker, J.C., Dummer, R., Hartmann, A.A., Burg, G. and Schmidt, R.E., 1991, Shedding of ICAM-1 from human melanoma cell lines induced by IFNg and TNFa. *J. Immunol.* 147:4398-4401.

Birdsall, H.H., Lane, C., Ramser, M.N. & Anderson, D.C., 1992, Induction of VCAM-1 and ICAM-1 on human neural cells and mechanisms of mononuclear leucocyte adherence. *J. Immunol.* 148:2717-2723.

Brown, Z., Neild, G.H., Willoughby, J.J., Somia, N.V. & Cameron, S.J., 1986, Increased factor VIII as an index of vascular injury in cyclosporine nephrotoxicity. *Transplantation.* 42:150-153.

Dunlop, L.C., Skinner, M.P., Bendall, L.J., Favaloro, E.J., Castaldi, P.A., Gorman, J.J., Gamble, J.R., Vadas, M.A. & Berndt, M.C., 1992, Characterization of GMP-140 (P-selectin) as a circulating plasma protein. *J. Exp. Med.* 175:1147-1150.

Faull, R.J. & Russ, G.R., 1989, Tubular expression of ICAM-1 during renal allograft rejection. *Transplantation.* 48:226-230.

Fortes, Z.B., Farsky, S.P., Oliviera, M.A. & Garcia-Leme, J., 1991, Direct vital microscopic study of defective leucocyte-endothelial interaction in diabetes mellitus. *Diabetes* 40:1267-1273.

Gearing, A.J.H., Hemmingway, I., Pigott, R., Hughes, J., Rees, A.J., Cashman, S.J., 1992, Soluble forms of vascular adhesion molecules, E-selectin, ICAM-1 and VCAM-1: Pathological significance. *Ann. N. Y. Acad. Sci.* 667:324-331.

Giavazzi, R., Chirivi, R.G.S., Garofalo, A., Rambaldi, A., Hemingway, I., Pigott, R. & Gearing, A.J.H., 1992, Soluble ICAM-1 is released by Human melanoma cells and is associated with tumor growth in nude mice. *Cancer Res.* 52:2628-2630.

Harning, R., Mainolfi, E., Bystryn, J-C., Henn, M., Merluzzi, V.J. and Rothlein, R., 1991, Serum levels of circulating intercellular adhesion molecule 1 in human malignant melanoma. *Cancer Res.* 51:5003-5005.

Johnston, G.I., Cook, R.G. and McEver, R.P., 1989, Cloning of GMP140, a granule membrane protein of platelets and endothelium: sequence similarity to proteins involved in cell adhesion and inflammation. *Cell.* 56: 1033-1044.

Johnston, G.I., Bliss, G.A., Newman, P.J. and McEver, R.P., 1990, Structure of the human gene encoding GMP-140, a member of the selectin family of adhesion receptors for leukocytes. *J. Biol. Chem.* 265:21381-21385.

Kishimoto, T.K., Jutila, M.A., Berg, E.L. and Butcher, E.C., 1989, Neutrophil MAC-1 and MEL-14 adhesion proteins inversely regulated by chemotactic factors. *Science* 245:1238.

Lawrence, M.B. and Springer, T.A., 1991, Leukocytes roll on a selectin at physiologic flow rates: distinction from and prerequisite for adhesion through integrins. *Cell* 65:859-873.

Leung, D.Y., Pober, J.S. and Cotran, R.S., 1991, Expression of ELAM-1 in elicited late phase allergic reactions. *J. Clin. Invest.* 87: 1805-1809.

Lo, S.K., Lee, S., Ramos, R.A., Lobb, R., Rosa, M., Chi-Rosso, G., Wright, S.D., 1991, E-Selectin stimulates the adhesive activity of leucocyte integrin CR3 (CD11b/CD18, Mac-1, amb2) on human neutrophils. *J. Exp. Med.* 173:1493-1500.

Marlin, S.D., Staunton, D.E., Springer, T.A., Stratowa, C., Sommergruber, W., Merluzzi, V.J., 1990, A soluble form of intercellular adhesion molecule-1 inhibits rhinovirus infection. *Nature* 344:702.

Mulligan, M.S., Varani, J., Dame, M.K., Lane, C.L., Smith, C.W., Anderson, D.C. and Ward, P.A., 1991, Role of ELAM-1 in neutrophil mediated lung injury in rats. *J. Clin. Invest.* 88: 1396--1406.

Norris, P., Poston, R.N., Thomas, D.S., Thornhill, M., Hawk, J. and Haskard, D.O., 1991, The expression of ELAM-1, ICAM-1 and VCAM-1 in experimental cutaneous inflammation: a comparison of UVb erythema and delayed hypersensitivity. *J. Invest. Dermatol.* 96:.

Osborn, L., Hesslon, C., Tizard, R., Vassalo, C., Luhowsky, S., Chi-Rosso, G., Lobb, R., 1989, Direct expression cloning of VCAM-1, a cytokine-inducible endothelial protein that binds to lymphocytes. *Cell* 59:1203-1211.

Pasi, K.J., Enayat, M.S., Horrocks, P.M., Wright, A.D., Hill, F.G., 1990, Qualitative and quantitative abnormalities of von Willebrand antigen in patients with diabetes mellitus. *Thromb. Res.* 59:5-81-91.

Pigott, R., Needham, L.A., Edwards, R.M., Walker, C.A. and Power, C., 1991, Structural and functional studies of the activation antigen ELAM-1 using a panel of monoclonal antibodies. *J. Immunol.* 147:130-135.

Pigott, R. Dillon, L.P., Hemingway, I. and Gearing, A.J.H., 1992, Soluble forms of E-selectin, ICAM-1 and VCAM-1 are present in the supernatants of cytoline-activated cultured endothelial cells. Biochem. *Biophys. Res. Comm.* 187:584-589.

Rice, G.E., Munro, J.M., Corless, C. and Bevilacqua, 1991, Vascular and non-vascular expression of INCAM-110. *Am. J. Pathol.* 138: 385-393.

Rothlein, R., Mainolfi, E.A., Czajkowski, M., and Marlin, S.D., 1991, A form of circulating ICAM-1 in human serum. *J. Immunol.* 147:378-3793.

Ruco, L.P., Pomponi, D., Pigott, R., Gearing, A.J.H., Baiocchini, A. and Baroni, C.D., 1992, Expression and cell distribution of the adhesion molecules ICAM-1, VCAM-1, ELAM-1 and endoCAM (CD31) in reactive human lymph nodes and in Hodgkins disease. *Am. J. Pathol* 140:1337-1344.

Ruco, L.P., Gearing, A.J.H., Pigott, R., Pomponi, D., Burgiov, D., Cafolla, A.A., Baiocchini, A. and Baroni, C.D., 1991, Expression of ICAM-1, VCAM-1 and ELAM-1 in angiofollicular lymph node hyperplasia (Castlemans disease) evidence for dysplasia of follicular dendritic reticulum cells. *Histopathology* 19:523-528.

Ryan, J., Benynon, H., Rees, A.J., Cassidy, M.J.D., 1991, Evaluation of the *in vitro* production of tumor necrosis factor by monocytes in dialysis patients. *Blood Purif* 9:142-147.

Seth R., Raymond F.D., and Makgoba M.W., 1991, Circulating ICAM-1 isoforms: diagnostic prospects for inflammatory and immune disorders. *Lancet* 338:83-84.

Simmons, D., Makgoba, M.W. and Seed, B., 1988, ICAM an adhesion ligand of LFA-1 is homologous to the neural cell adhesion molecule NCAM. *Nature.* 331: 624-627.

Springer, T.A., 1990, Adhesion receptors of the immune system. *Nature.* 346: 425-434.

Tsujisaki, M., Hirata, I.H., Hanzawa, Y., Masuya, J., Nakano, T., Sugiyama, T., Matsui, M., Hinoda, Y. and Yachi, A., 1991, Detection of circulating ICAM-1 antigen in malignant disease. *Clin. Exp. Immunol.* 85:3-8.

Van Seventer, G.A., Newman, W., Shimizu, Y., Nutman, T.B., Tanaka, Y., Horgan, K.J., Gopal, T.V., Ennis, E., O'Sullivan, D., Grey, H. and Shaw, S., 1991, Analysis of T cell stimulation by superantigen plus MHC class II molecules or by CD3 monoclonals: costimulation by purified adhesion ligands VCAM-1, ICAM-1 but not ELAM-1. *J. Exp. Med.* 174:901-913.

Warrens, A.N., Cassidy, M.J.D., Takahashi, K., Ghatei, M.A., Bloom, S.R., 1990, Endothelin in renal failure. *Nephrol. Dial. Transplant.* 5:418-422.

Watson, S.R., Fennie, C. and Lasky, L.A., 1991, Neutrophil influx into an inflammatory site inhibited by a soluble homing receptor chimaera. *Nature* 349:164-167.

Wegner, C.D., Gundel, R.H., Reilly, P., Haynes, L.G. and Rothlein, R., 1990, ICAM-1 in the pathogenesis of asthma. *Science* 247: 456-459.

V.   ORGAN TRANSPLANTATION/REJECTION

EXPERIMENTAL CELL TRANSPLANTATION AND GENE THERAPY TECHNIQUES IN THE TREATMENT OF LIVER DISEASE

Achilles A. Demetriou, J. Rozga, D. Neuzil,* M. Holzman,
* D. Griffin,* Albert D. Moscioni

Cedars-Sinai Medical Center
Department of Surgery
Los Angeles, California
U.S.A.

and

Vanderbilt University
School of Medicine
Department of Surgery*
Nashville, Tennessee
U.S.A.

Whole organ transplantation remains the only clinically effective method of treating acute and chronic liver failure due to specific genetic and other defects of liver function. However, there are several factors which limit the application of liver transplantation in the treatment of liver insufficiency: high cost, relatively high morbidity and limited organ donor availability. As a result, investigators have attempted to develop alternative methods to treat liver insufficiency due to genetic defects of liver function by using either isolated normal hepatocyte transplantation and/or *in vitro* and *in vivo* gene therapy techniques.

Several general approaches have been taken in developing these alternative experimental therapeutic strategies:

A. To transplant normal allogeneic hepatocytes (or other cells) into an immunosuppressed recipient with a specific genetic liver defect.

B. To remove hepatocytes (or other cells) from a donor with a specific liver defect, genetically manipulate them *in vivo*, expand the corrected cell population and re-transplant them into the original donor without need for immunosuppression.

C. To correct the underlying liver defect by directly delivering the missing (or corrected gene) into the host hepatocytes *in vivo*.

We have carried out experiments, briefly described below, utilizing these strategies.

Vascular Endothelium, Edited by J.D. Catravas *et al.*
Plenum Press, New York, 1993

ALLOGENEIC CELL TRANSPLANTATION

Potential advantages of use of hepatocyte transplantation include: cell cryopreservation, which allows use of cells on demand; *in vitro* cell manipulation to decrease immunogenicity; use of cells from a single donor to treat multiple recipients and technically simple therapy at potentially low cost.

Transplantation of hepatocytes, injection of various hepatocyte extracts and hepatocyte culture supernatants, have been shown to prolong survival in animals with D(+)galactosamine-induced liver injury and animals with acute liver ischemia (*Baumgartner et al., 1983; Sommer et al., 1980*). Hepatocyte transplantation into the spleen, fat pads, dorsal fascia, pancreas and portal vein, has been carried out resulting in partial correction of specific liver metabolic defects and demonstration of viable differentiated hepatocytes at various ectopic sites (*Mito et al., 1978*).

Our laboratory has been working towards developing techniques of experimental isolated hepatocyte transplantation. Specifically, we have described techniques of intraperitoneal cell transplantation utilizing a collagen matrix and selective intraportal transplantation (*Demetriou et al., 1986a; Holzman et al., 1992*). These studies are summarized below:

INTRAPERITONEAL TRANSPLANTATION

Correction of Specific Genetic Liver Function Defects

Normal rat hepatocytes were harvested by collagenase portal vein perfusion and were attached to type-l collagen-coated dextran microcarriers (*Demetriou et al., 1986a*). Microcarrier-attached normal rat hepatocytes were then transplanted into animals with specific genetic defects of liver function: rats unable to conjugate bilirubin (Gunn rats) and synthesize albumin (Nagase analbuminemic rats; NAR). We were able to demonstrate survival and function of transplanted normal hepatocytes in immunosuppressed recipients by: a) the appearance of bilirubin conjugates, increase in total bilirubin in the bile and decrease in serum bilirubin levels in transplanted Gunn rats and b) an increase in plasma albumin levels in NAR recipients (*Demetriou et al., 1986a; Demetriou et al., 1986b*).

Use of Allogeneic Normal Hepatocytes to Treat Acute Liver Insufficiency

Microcarrier-attached normal rat hepatocytes were transplanted into rats with acute liver insufficiency following 90% partial hepatectomy. A dramatic improvement in recipient survival and prevention of post-operative hypoglycemia in rats transplanted four days prior to 90% partial hepatectomy was noted (*Demetriou et al., 1988*).

Transplanted Hepatocyte Proliferation

In the experiments in which animals undergoing 90% partial hepatectomy demonstrated improved survival following hepatocyte transplantation, it was noted that transplantation of only a small number of liver cells (approximately 1-2% of the total liver mass), resulted in improvement in animal survival (*Demetriou et al., 1988*). One possible explanation is that transplanted hepatocytes proliferated at their ectopic site. Studies were undertaken to determine whether transplanted microcarrier-attached hepatocytes proliferate in the peritoneal cavity of syngeneic rat recipients. Microcarrier-attached normal hepatocytes were transplanted intraperitoneally into syngeneic rats. Three days later, transplanted rats underwent 70% partial hepatectomy and 24 hours postoperatively, they were injected with [³H] thymidine intravenously. One hour later, all rats were killed with ether and their livers and peritoneal cavity aggregates were excised and assayed for total DNA and [³H] thymidine incorporation into DNA. As expected, there was a significant increase in [³H] thymidine incorporation into DNA in the regenerating liver remnants; similarly, there was a significant increase in [³H]thymidine uptake by peritoneal aggregates in partially hepatectomized rats (*Arepally et al., 1987*). However, using either [³H] thymidine autoradiography or staining for bromodeoxyuridine, we demonstrated that although cells within the

126

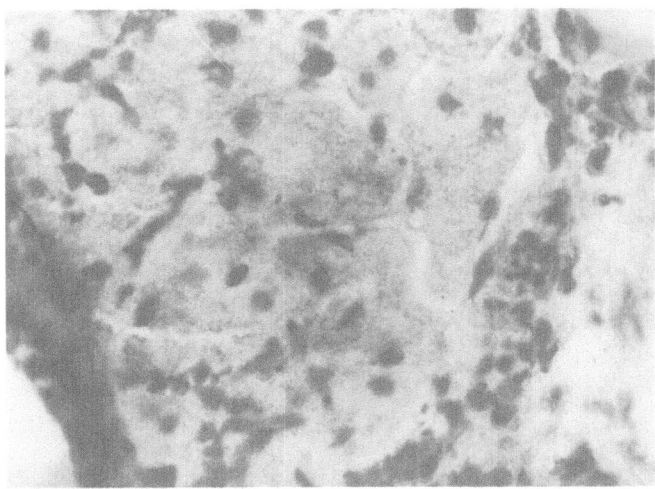

FIGURE 1: Hepatocytes staining for albumin (brown color) using a specific anti-rat albumin antibody are shown 28 days following transplantation into the peritoneal cavity. Most hepatocytes are present in the space among microcarriers rather than on the edge of the photomicrograph.

peritoneal cavity aggregates proliferated in response to partial hepatectomy, most proliferating cells were not differentiated hepatocytes. Thus, we have not been able to convincingly demonstrate hepatocyte proliferation at this ectopic site.

MORPHOLOGIC STUDIES

Intraperitoneally transplanted microcarrier-attached hepatocytes in rats formed well-vascularized aggregates on the anterior surface of the pancreas. In the aggregates, microcarriers were surrounded by *epithelioid* cells. Using indirect immunoperoxidase methods, formalin-fixed, paraffin-embedded sections of peritoneal cavity aggregates were stained with anti-rat albumin antibody and many hepatocytes were identified staining for albumin up to 28 days following transplantation in cyclosporin-treated recipients (Figure 1). Electron microscopic examination of the aggregates revealed formation of microvilli and canaliculi and tight junctions suggesting a high level of morphologic differentiation (Figure 2).

MODULATION OF TRANSPLANTED CELL ANTIGENICITY

We examined the effect of ultraviolet (UV) irradiation on the survival and function of normal hepatocytes *in vivo*. Microcarrier-attached normal rat hepatocytes were pretreated with UV irradiation ($600J/m^2$) prior to transplantation into allogeneic NAR recipients. Transplanted hepatocyte survival and function were assessed by serial measurements of plasma albumin levels. There was a significantly prolonged elevation of plasma albumin levels in rats transplanted with UV-treated cells when compared to rats transplanted with sham-irradiated cells (*Phay et al., 1988*).

Cell encapsulation has been shown to prolong transplanted cell survival and function because the semipermeable membrane reportedly immunoisolates the cells. Employing 25,000 MW alginate/poly-L-lysine (*ALP; Cai et al., 1988*), we carried out a series of experiments using encapsulated microcarrier-attached hepatocytes. NAR rats were transplanted either with microcarrier-attached allogeneic hepatocytes or with encapsulated microcarrier-attached hepatocytes. No statistically significant differences

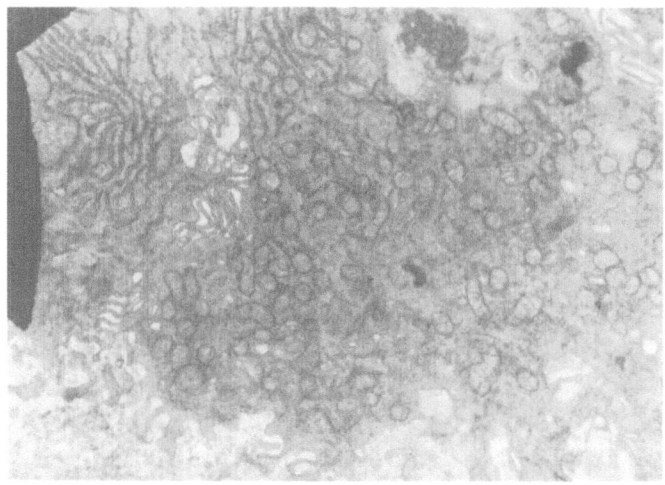

FIGURE 2: Transmission electron micrograph of a section through the peritoneal cavity cell / microcarrier aggregate, depicting hepatocytes forming canaliculi with microvilli and areas which appear similar to tight junctions.

between the two groups were noted (*Rozga et al., 1991*). A very strong inflammatory response was seen in the peritoneal cavity of animals transplanted with encapsulated microcarrier-attached hepatocytes. No such reaction was seen in controls injected with encapsulated microcarriers alone or animals transplanted with syngeneic microcarrier--attached encapsulated hepatocytes, suggesting that allogeneic hepatocytes were indeed recognized as foreign and induced a significant pericapsular inflammatory reaction, probably due to release of cytokines and other inflammatory mediators.

INTRAPORTAL HEPATOCYTE TRANSPLANTATION

Intraportal hepatocyte transplantation has not been widely used because of development of portal hypertension and multifocal necrosis when large numbers of cells were given. We have developed a model of selective intraportal infusion of hepatocytes into some liver lobes, but not others, which allows portal decompression following transplantation (*Holzman et al., 1992*). Recipient NAR rats underwent cannulation of the gastroduodenal vein and hepatocyte (2×10^7) infusion. Either the anterior or posterior liver lobes were selectively infused with normal hepatocytes by occluding the vascular supply of the non-perfused lobes during the infusion. Controls were infused with NAR hepatocytes. All recipients were treated with cyclosporin for the duration of the experiment. Animals transplanted with normal hepatocytes demonstrated sustained significant (1,500%) elevation in plasma albumin levels up to 56 days post-transplantation (duration of the experiment); clusters of albumin-producing hepatocytes were detected in liver sections stained immunohistochemically (Figure 3). No significant changes in plasma albumin levels were noted in control NAR recipients transplanted with NAR hepatocytes, and no hepatocyte clusters could be detected immunohistochemically in liver sections from these animals (*Holzman et al., 1992*). Our results demonstrate that selective intraportal infusion of hepatocytes is an effective technique of hepatocyte transplantation.

Although we have made significant progress in the field of hepatocyte transplantation, several important questions and issues need to be resolved before serious clinical trials of hepatocyte transplantation in humans can be carried out with a realistic chance of success. These include: determination of the optimal site for hepatocyte transplantation, need for accurate data concerning the minimum number of cells needed to provide metabolic support and carry out specific liver functions,

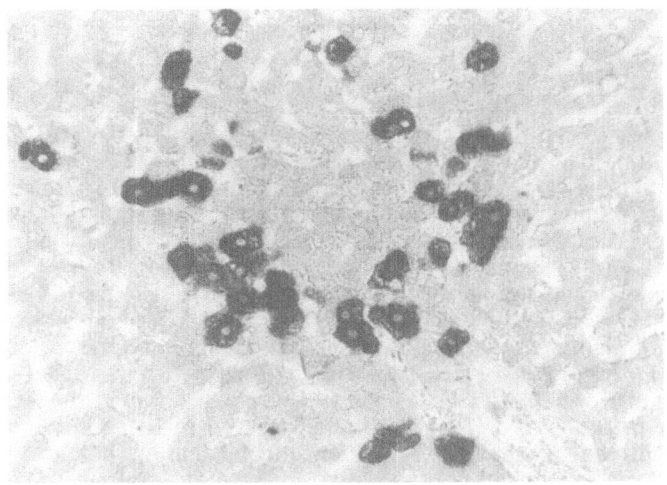

FIGURE 3: A cluster of transplanted normal hepatocyte staining positively for albumin in the liver parenchyma of an NAR rat, days following selective intraportal transplantation.

failure to achieve indefinite successful metabolic support following hepatocyte transplantation even in syngeneic animals, lack of knowledge concerning the optimal type of immunosuppression needed following allogeneic hepatocyte transplantation and lack of techniques to detect transplanted cell rejection and inability to demonstrate significant levels of hepatocyte proliferation either *in vitro* or *in vivo*.

TRANSPLANTATION OF CORRECTED CELLS

Hepatocytes

Initial experiments were carried out to develop reproducible methods of normal adult hepatocyte transduction and expression of retrovirally-introduced genes. Two retroviral vectors were tested: an ecotropic virus expressing β-galactosidase (β-gal) from E. coli and an amphotropic virus expressing the bacterial gene conferring resistance to Hygromycin-B (HB) in eukaryotic cells. Viral titers were determined in NIH 3T3 cells. Hepatocytes were isolated from adult male Wistar rats by collagenase perfusion. Isolated hepatocytes were plated onto collagen-coated tissue culture plates (1×10^6 cells/35 mm plate) and cultured in hormonally defined serum free media (RPMI 1640). After 48 hours, attached cells were exposed to a retroviral supernatant (1×10^6 virions/ml; 3ml/plate) for a 24-hour period. Following transduction, retroviral media were replaced with normal media and cultures were continued until examined for viral gene expression. Four days post-transduction with the ecotropic virus, hepatocyte cultures were tested for β-gal expression using X-Gal (Sigma) as substrate. Positive cells stained blue. Twenty-four hours following transduction with the amphotropic virus, hepatocyte cultures were tested for HB resistance by exposure to media containing HB (Sigma; 55 μg/ml, 3ml); four days later, culture plates were fixed and stained with Giemsa to highlight viable cell colonies. At seven days, control hepatocyte cultures demonstrated no β-gal activity, whereas, hepatocyte cultures transduced with the ecotropic virus, had a small number of cells ($1 \pm 0.5\%$) expressing β-gal activity. Control hepatocyte cultures exposed to HB had $5 \pm 5\%$ survival at four days. Cultures exposed to the amphotropic virus and tested for HB had $60 \pm 10\%$ survival ($P < 0.01$).

Transduced HB-resistant hepatocytes were recovered and transplanted into syngeneic rats by either attaching them to microcarriers and injecting them into the peritoneal cavity, as described above, or selectively infusing them into the portal vein. However, only a negligible number of transduced cells could be detected *in vivo*.

Fibroblasts

In vitro **studies.** Human and rat fibroblasts were harvested using primary explant techniques. Human fibroblasts were obtained from normal neonates undergoing circumcision. Skin was harvested from the dorsum of rats. Two retroviral vectors employed for this study were provided by Drs. Robert Overell and Steven Lupton of the Immunex Corp (Seattle, WA). Both vectors were constructed with double promoters for control of expression of inserted foreign genes. Both vectors contained the amphotropic ᴧ(-)gH retroviral vector which confers resistance to HB. This vector is constructed from ᴧH and contains the HB resistance *hph* gene, which is expressed under control of the human cytomegalovirus (CMV) promoter. One vector, in addition, contained the *lacZ* gene conferring expression of β-gal, which was under control of the viral long terminal repeat (LTR). The second vector contained the *hF-IX* similarly under control of the viral LTR. Fibroblasts, cultured in tissue culture flasks, were transduced when they reached 50% confluence by incubating them overnight in viral supernatant media and polybrene (4 μg/ml). After transduction, cells were placed in normal culture media. Clones developed from transduced fibroblasts were maintained in selection media for the duration of the experiment.

Factor IX was assayed in culture media using an enzyme-linked immuno-absorbance assay (ELISA), provided in a kit from Diagnostica (Asserachrom IX: Antigen; Stago, France). Popymerase chain reaction (PCR) amplification was performed on genomic DNA isolated from cultured transduced fibroblasts. In positive samples, the primers generated a 975 bp fragment, which was then separated by agarose gel electrophoresis and identified by co-chromatography with a known *hph* gene fragment following ethidium bromide staining.

Selected HB-resistant fibroblast clones (rat and human) demonstrated significant hF-IX production; control non-transduced fibroblasts showed no production of hF-IX (*Neuzil et al., 1992*). Similarly, factor IX clotting activity, measured by the one-stage clotting assay, could be demonstrated both in transduced rat and human fibroblasts (*Neuzil et al., 1992*). All transduced cell lines retained their ability to produce significant quantities of hF-lX *in vitro* for up to six months following transduction (*Neuzil et al., 1992*). Storage at -70 °C for two months did not have an effect on *in vitro* hF-IX production by the transduced fibroblasts. PCR was used to demonstrate incorporation of the proviral DNA sequence into fibroblast DNA. Transduced fibroblasts exhibited incorporation of the HB resistance proviral sequence into their genome. Southern blot analysis confirmed the PCR data (*Neuzil et al., 1992*).

In Vivo **Studies.** Transduced and non-transduced syngeneic Lewis rat fibroblasts were harvested from tissue culture plates and resuspended in heparinized saline (25 U/ml). Through a mid-line abdominal incision, under ether anesthesia, syngeneic hosts were transplanted either by a single intraportal injection of a large number of cells in suspension (2×10^7 cells/3 ml) or by injecting the same number of cells using multiple injections *via* pediatric catheter (5mm Port-A-Cath, Strato-Medical Corp. Beverly, MA) into the portal vein. Additionally, rats underwent intraperitoneal transplantation of either control or transduced HB-resistant hF-IX syngeneic fibroblasts attached to collagen-coated dextran microcarriers (1-2 $\times 10^7$ cells/0.2g dry weight microcarriers). At one and four weeks after transplantation, cell-transplanted livers and peritoneal cavity microcarrier/cell aggregates, as well as tissues from control animals, were collected. Specimens were processed for Hematoxylin/Eosin or immunohistochemical staining for hF-IX using a monoclonal rabbit anti-hF-IX antibody and a Vectastain ABC Kit (Vector Laboratories, Burlingame, CA) employing a peroxidase-coupled secondary antibody and diaminobenzidine as substrate. Liver sections were examined and positively-stained cells counted. In sections from peritoneal cavity cell/microcarrier aggregates, the number of microcarriers per field was determined and the number of cells attached to each microcarrier was counted.

In the experimental animals transplanted with cells into the portal vein either as a single injection or in the form of repeated infusions through an indwelling catheter, no transduced transplanted cells could be identified using immunohistochemi-

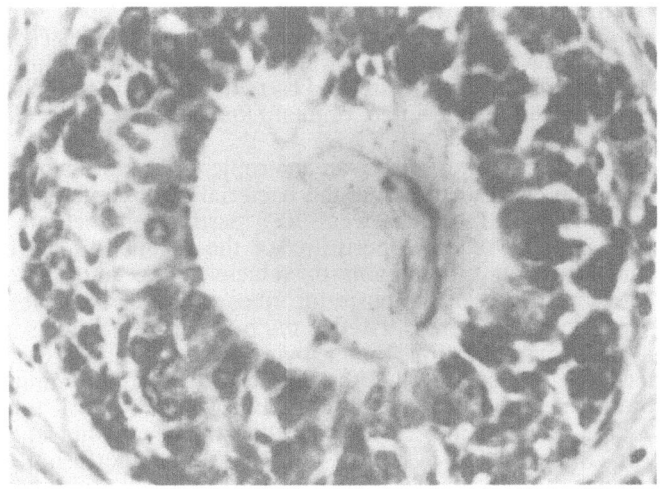

FIGURE 4: Human factor IX-stained fibroblasts on the surface of a microcarrier days following transplantation into a syngeneic Lewis rat; the cells appear more columnar and epithelioid.

cal techniques, and no hF-IX could be detected in their plasma (*Neuzil et al., 1992*). Following intraperitoneal transplantation, cells and microcarriers formed large aggregates on the anterior surface of the pancreas. Sections through these aggregates stained immunohistochemically demonstrated a large number of cells expressing hF-IX (Figure 4). At one and two weeks after transplantation, a significant number of hF-IX-stained cells was detected (95.6±8.1 and 125.5±14.1 positively-stained cells/microcarrier at one and two weeks, respectively). Positive cytoplasmic staining for hF-IX was seen in all experimental animals, and no staining was seen in any of the control animals up to eight weeks post transplantation, the time of termination of the experiment (*Neuzil et al., 1992*). It is of interest to note that transduced, transplanted cells attached to microcarriers assumed a columnar shape similar to that of most secretory, or epithelioid, cells, while non-transduced transplanted fibroblasts remained flattened on the microcarrier surface. In this experiment, we were again unable to demonstrate an increase in hF-IX plasma levels in any of the transplanted recipients (*Neuzil et al., 1992*).

Our results demonstrate that rat and human fibroblasts can be successfully transduced with a retrovirus containing the hF-IX gene, and this can result in significant Factor IX production *in vitro*. Following intraperitoneal transplantation of microcarrier-attached, transduced rat fibroblasts, large numbers of viable cells expressing hF-IX could be identified for the duration of the experiment using immunohistochemical techniques. However, no increase in Factor IX plasma levels could be demonstrated. This may be due to hF-IX destruction by rat serum proteases, inadequate level of hF-IX production by transplanted cells, transplantation of an inadequate number of cells, formation of anti-hF-IX antibodies, and suppression of hF-IX production by the normal rat recipients, which have normal levels of endogenously produced rat clotting factors. Neither a single intraportal injection of fibroblasts nor repeated injections had any demonstratable effects.

IN VIVO GENE THERAPY TECHNIQUES

Recently, several investigators succeeded in delivering a functional gene to hepatocytes *in vitro*, but a number of formidable problems remain to be resolved to achieve stable, long-term expression of the genes. These include: improved transduction efficiency, expansion of the population of transduced cells and development of

131

techniques of re-transplantation in the original donor. Attempts to deliver genes directly to cells *in vivo* have been reported with short-term success (*Wu et al., 1989; Wolff et al., 1990*). We have developed a novel technique of selective *in situ* perfusion of the rat liver, which was successfully used to carry out retrovirus-mediated transduction of hepatocytes *in vivo* in the intact animal (*Rozga et al., 1992; Moscioni et al., 1992*).

Two retroviral vectors were used: an ecotropic virus expressing β-gal from E. coli and an amphotropic virus expressing the bacterial gene conferring resistance to HB. Adult Sprague-Dawley rats underwent 70% partial hepatectomy under ether, followed 20 hours later by selective perfusion of the regenerating liver lobes with either one of the retroviral supernatants or vehicle alone. Another group of rats underwent selective perfusion of the posterior lobes without partial hepatectomy. In all groups, the posterior lobes were isolated *in situ* by inserting cannulas for inflow and outflow control under magnification. Either a viral supernatant (1×10^6 virions/ml) or vehicle was selectively perfused through the posterior lobes for 15 minutes; following perfusion, normal liver blood flow was established by removing all tourniquets and cannulae. At 7 days, rat hepatocytes transduced with the amphotropic virus, were isolated from the liver remnant by portal vein collagenase perfusion and cultured. Cells were selected with HB (LD99 = 500 μg/ml) and cell viability was assessed following fixation and Giemsa staining. For the ecotropic β-gal virus, the liver was perfused with fixative and multiple sections stained with X-gal to identify enzyme expression. Partially hepatectomized rats demonstrated significant levels of *in vivo* hepatocyte transduction with both retroviruses; ecotropic (β-gal): 5±2% (compared to 0% in controls, p < 0.001), amphotropic (HB-resistance): 26±9% (compared to 1±0.5% in controls, P < 0.001). No significant levels of transduction were noted in all non-hepatectomized rats. It appears that hepatocyte proliferation is needed for efficient *in vivo* liver cell transduction. A novel surgical procedure is described which permits targeted delivery of retrovirally-mediated genes to the liver *in vivo* (*Rozga et al., 1992; Moscioni et al., 1992*).

ACKNOWLEDGEMENTS

The authors thank Drs. Robert Overell and Steven Lupton of Immunex Corporation (Seattle, WA) for providing the retroviral vectors employed in this study and for their technical advice.

REFERENCES

Arepally, G., Felcher, A., Moscioni, A.D., Demetriou, A.A., 1988, Proliferation of intraperitoneally transplanted microcarrier-attached liver cells. *Surg. Forum.* 39:396.

Baumgartner, D., LaPlante-O'Neill, P.M., Sutherland, D.E., Najarian, J.S., 1983, Effect of intrasplenic injection of hepatocytes, hepatocyte fragments and hepatocyte culture supernatants on D-galacctosamine-induced liver failure. *Eur. J. Surg. Res.* 15:129.

Cai, Z., Shi, Z., O'Shea, G.M., Sun, A.M., 1988, Microencapsulated hepatocytes for bioartificial liver support. *Artificial Organs*, 12:388.

Demetriou, A.A., Whiting, J., Feldman, D., Levenson, S.M., Moscioni, A.D., Kram, M., Chowdhury, N.R., Chowdhury, J.R., 1986a, Replacement of liver function in rats with specific inherited metabolic errors by transplantation of microcarrier-attached hepatocytes. *Science*, 233:1190.

Demetriou, A.A., Whiting, J., Feldman, D., Levenson, S.M., Chowdhury, J.R., 1986b, New method of hepatocyte transplantation and extracorporeal liver support. *Ann. Surg.*, 204:259.

Demetriou, A.S., Reisner, A., Sanchez, J., Levenson, S.M., Moscioni, A.D., Chowdhury, J.R., 1988, Transplantation of microcarrier-attached hepatocytes into 90%. partially hepatectomized rats. *Hepatology*, 8:1006.

Holzman, M.D., Rozga, J., Neuzil, D., Griffin, D., Moscioni, A.D., Demetriou, A.A. Intraportal hepatocyte transplantation and expression of albumin synthesis *in vivo* in analbuminemic rats. *Transplantation*, (Submitted).

Mito, M., Kusano, M., Onishi, T., Saito, T., Ebata, H., 1978, Hepatocellular transplantation. *Gastroenterol.* Japan 13:480.

Moscioni, A.D., Rozga, J., Neuzil, D., Overrel, R.W., Holt, J.T., Demetriou, A.A. *In vivo* regional deli
very of retrovirally-mediated foreign genes to rat liver cells: Need for partial hepatectomy for
successful foreign gene expression. *Surgery,* (In Press).

Neuzil, D., Holzman, M., Rozga, J., Griffin, D., Moscioni, A.D., Demetriou, A.A. Transplantation of
transduced fibroblasts in rats; function of retrovirally-mediated foreign genes. *J. Surg. Res.*,
(Submitted).

Phay, J., Felcher, A., Levenson, S.M., Demetriou, A.A., 1988, Prolongation of function of transplanted
microcarrier-attached hepatocytes by ultraviolet irradiation. *Surg. Forum*, 39:398.

Rozga, J., Bellew, T., Moscioni, A.D., Williams, R.F., Demetriou, A.A., 1991, Transplantation of
microencapsulated hepatocytes into analbuminemic rats. *Surg. Forum*, 42:419.

Rozga, J., Moscioni, A.D., Neuzil, D., Demetriou, A.A., 1992, A model for directed foreign gene delivery to rat liver cells *in vivo. J. Surg. Res.*, 52:209.

Sommer, B.G., Sutherland, D.E., Simmons, R.L., Najarian, J.S., 1980, Hepatocellular transplantation
for experimental ischemic acute liver failure in dogs. *J. Surg. Res.*, 29:319.

Wolff, J.A., Malone, R.W., Williams, P., Chong, W., Acsadi, G., Jani, A., Felgner, P.L., 1990, Direct
gene transfer into mouse muscle *in vivo. Science*, 247:1465.

Wu, C.H., Wilson, J.M., Wu, G.Y., 1989, Targeting genes: Delivery and persistent expression of a foreign gene driven by mammalian regulatory elements *in vivo. J. Biol. Chem.*, 264:16985.

MODULATORS OF CHRONIC REJECTION

*Bengt Fellström and **Erik Larsson

*Department of Internal Medicine
University Hospital
Uppsala
Sweden

**Department of Pathology
University Hospital
Uppsala
Sweden

INTRODUCTION

Chronic rejection is one of the major threats of graft function on a long-term basis in heart and kidney transplantation. During the last decades, the results of organ transplantation have improved steadily, whereas the annual rate of graft loss after the first post-transplantation year has not changed significantly (*Cook and Terasaki, 1989; Thorogood et al., 1992*). The diagnosis of chronic rejection is based on a combination of clinical, morphological and angiographic findings. The histopathology of chronic renal transplant rejection is characterized by various degrees of narrowing of the graft arteries and arterioles, interstitial cellular infiltration and fibrosis, tubular atrophy and variable glomerular changes (*Maryniak et al., 1985; Kasiske et al., 1991; Paul et al., 1992*). The vascular narrowing results from infiltration of the intima by mononuclear cells, migration and proliferation of vascular smooth muscle cells and fibroblasts from the media into the intima and subsequent deposition of extracellular matrix material. The histopathological picture of chronic heart graft rejection is characterized by atherosclerotic vascular changes in combination with variable degrees of interstitial fibrosis (*Gao et al., 1987; Cary, 1992*). The atherosclerotic vessel wall lesions are important features of chronic rejection in hearts and kidneys, but not specific, since they may also be observed in other vascularized organ allografts.

It has been estimated that renal graft failure due to chronic rejection is the leading cause of 6-12% of graft losses observed after the initial 5-10 years following transplantation (*Mahony and Sheil, 1987; Paul, 1992*). It has also been estimated that 25% of the cumulated graft losses in the first three posttransplant years were due to chronic rejection in transplanted hearts. Systematic annual coronary angiographic studies of cardiac allografts suggest that 40-60% of the grafts have significant vascular changes within 5 years following transplantation, indicative of chronic rejection (*Paul, 1992*).

PATHOPHYSIOLOGY

The pathophysiology of chronic rejection is complex, and it is difficult to pin-point a single leading cause. Notwithstanding, morphologic and functional

Development of Chronic Rejection

Progression of Vascular Arteriopathy

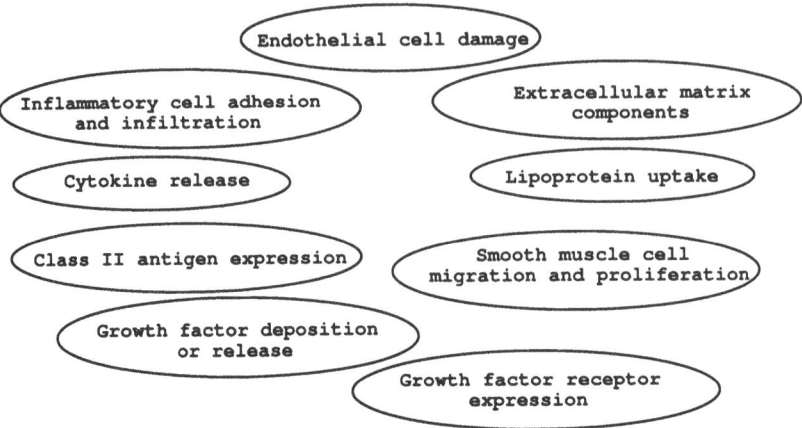

FIGURE 1: Factors of pathogenetic importance for the progression of transplant atherosclerosis.

similarities between the vessel wall lesions of graft atherosclerosis and *naturally occurring* atherosclerosis, it may be assumed that the former is a manifestation of atherosclerosis partially driven by allogeneic immune mechanisms. The pathogenesis involves a number of mechanisms which have been demonstrated to be involved in the progression of the arteriopathy (Fig 1). Even though the endothelium may remain intact in a vessel undergoing atherosclerotic transformation, early *endothelial cell damage* is often present and may be due to a number of causes. In transplantation, such possible causes are ischemic and reperfusion damage (*Ross, 1986; Lehr et al., 1992*), the presence of the endothelial cell antibodies (*Dunn et al., 1992*) followed by complement activation. Specific enzymes such as heparinases (*Naparstek et al., 1984*) may also be present as well as deposition of LDL cholesterol or oxidatively modified lipoproteins (*Ross, 1986; Steinberg, 1989*).

The initial endothelial cell damage may be followed by monocyte or macrophage influx into the subendothelial compartment and deposition of platelet granule proteins. The endothelial cell will react to the damage and to release of cytokines, including interferon-γ and interleukin-1, by an increased expression of adhesion molecules (*Patarroyo et al., 1990; Springer, 1990; Hansson et al., 1992, Taylor et al., 1992*), upregulation of class-II antigen expression and increased permeability to plasma proteins. Both endothelial and smooth muscle cells may become activated during the vessel wall injury and produce multifunctional cytokines and growth factors which may induce cell proliferation and migration through paracrine, autocrine or juxtacrine pathways.

The list of *cytokines and growth factors* possibly involved in tissue remodelling of chronic rejection is long. There is evidence that platelet-derived growth factor (PDGF) plays a similar role in the development of chronic rejection as in naturally occurring atherosclerosis. PDGF may be released and deposited when platelets aggregate or become activated in the vascular wall, and it may also be secreted by a number of cells including monocytes and macrophages (*Shimokado et al., 1985*), endothelial cells (*DiCorleto and Bowen-Pope, 1983*), vascular smooth muscle cells (*Nilsson et al., 1985*) and glomerular mesangial cells (*Silver et al., 1989*). PBGF has also been demonstrated to be present in renal transplants undergoing chronic rejection as well as in experimental heart transplants in the rat, simulating chronic rejection.

Furthermore, it has been demonstrated that PDGF β-receptors may be upregulated in tissues going through a chronic inflammatory process, including

- Ox-LDL increases adhesion of monocytes to endothelial cells

- Ox-LDL causes release of a mononuclear leukocyte factor,
  which stimulates e.c. expression of
      Intracellular Adhesion Molecule-1 (ICAM-1, CD54)
      Vascular Cell Adhesion Molecule-1 (VCAM-1)
      Endothelial Selectin (E-Selectin, ELAM-1)

- Ox-LDL enhances class II antigen expression on human monocytes

- Ox-LDL stimulates the release of IL-1ß from monocytes

- Ox-LDL stimulates differentiation of monocytes to
  resident macrophages (LeuM3 ↑ , CD4 ↓)

- Ox-LDL increases DNA-synthesis, expression of HLA-DR and IL-2
  receptors on T-lymphocyte (activation of T-lymphocytes)

- Ox-LDL increases the expression of PDGF-AA transcripts in
  smooth muscle cells (SMC)

- Ox-LDL increases the expression of PDGF-a and ß receptors on SMC
  and the responsiveness of SMC to exogenous PDGF

FIGURE 2: Effects of oxidatively modified LDL on factors potentially important for atherosclerosis formation (adapted from the work of *Frostegård et al., 1990*).

rheumatoid arthritis (*Rubin et al., 1988*) and kidneys with chronic vascular rejection (*Fellstrom et al., 1989*). Similar increased receptor expression of PDGF α-receptors were found on smooth muscle cells in experimental heart and aorta transplants in the rat (unpublished observations).

Graft arteriosclerosis is usually more diffuse and wide spread than a normally occurring atherosclerosis, possibly secondary to that it is initiated and possibly propelled by the allogeneic immune reaction in the graft vessel wall. Subsequent vasculitis and activation of vascular wall cells may result in local production of growth promoting substances along with those substances infiltrating inflammatory cells, such as γ-interferon, IL-1, TNF-α, and TGF-β produced by T-cells. TGF-β may be of particular interest since it is known to modulate the immune or inflammatory reaction since it has a potential of either stimulating or inhibiting growth and differentiation of immune cells and mesenchymal cells. TGF-β may also stimulate the synthesis of extracellular matrix proteins and production of connective tissues (*Ruscetti and Palladino, 1991; Heldin and Miyazone, 1992*). Using monoclonal antibodies recognizing TGF-β1 and the latent TGF-binding protein (LTBP), it was demonstrated that these structures are present in experimental transplant atherosclerosis in both hearts and vessels (*Waltenberger et al., 1992*). Extraction of mRNA from these tissues followed by northern blot analysis demonstrated that it was mainly TGF-β1 present and TGF-β1 activity could also be demonstrated from extracted proteins from these tissues (*Waltenberger et al., 1992*).

Migration of smooth muscle cells into the proliferating intima and proliferation of smooth muscle cells are also a prominent feature in the development of atherosclerosis. This is also the case in experimental models of transplant arteriosclerosis and could be demonstrated by immunohistochemistry using antibodies against α-actine (*Wanders et al., 1992*).

Hyperlipoproteinemia has an established role in the development of naturally occurring atherosclerosis (*Ross, 1986; Steinberg, 1989; Regnström et al., 1992*). In particular, there is a correlation between hypercholesterolemia and the incidence of atherosclerotic lesions. It has been demonstrated that renal transplant patients with chronic vascular rejection have increased lipoprotein levels compared with patients with a stable graft function (*Dimény et al., 1992*). In particular, it could be demonstrated that patients with chronic rejection had an atherogenic pattern of lipoproteins and that there was a correlation between the degree of hyperlipidemia and the extent of pathological changes indicative of chronic rejection in graft biopsies. Furthermore, development of intimal hyperplasia in transplanted hearts in the rat were accelerated when the animals were fed a cholesterol enriched diet (*Fellström et al., 1990*). Since the uptake of lipoproteins in the vascular wall is associated with an influx of monocytes, which ingest LDL particles and become activated, a subsequent release of growth substances such as PDGF-like β peptides is a possible link between lipoproteins and induction of cell proliferation (*Ross, 1986*).

In the vascular wall of grafts going through chronic vascular rejection, there is almost constantly a more or less pronounced infiltration of T-lymphocytes and macrophages (*Fellström et al., 1988*). It may be assumed that the *oxidative modification of LDL cholesterol* is enhanced in cases with chronic vascular rejection. Based on the studies by the group at Karolinska Institute, oxidatively modified LDL may be an important inducer of changes in the vascular wall accelerating the atherosclerotic process. It was demonstrated that oxidized LDL causes a release of a mononuclear leukocyte factor, which stimulates the expression of ICAM-1, VCAM-1 and ILAM-1, leading to an increased adhesion of monocytes to endothelial cells (*Frostegard et al., 1991*). Furthermore, oxidized LDL may increase class-II antigen expression on human monocytes (*Frostegard et al., 1990*) and increase the expression of PDGF-α chain transcripts in smooth muscle cells and the expression of both α-and β-receptors on smooth muscle cells, as well as a subsequent enhanced responsiveness of SMC to exogenous PDGF (*Nilsson, 1992*) (see Fig 2). This is the background for trials with antioxidant agents such as Probucol[*], which has been initiated in chronic rejection.

Prostaglandins, prostacyclins and thromboxane have been recognized to be important mediators of various biological processes and also shown to be disturbed in chronic rejection. An imbalance has been demonstrated with high thromboxane levels and normal or slightly decreased prostacyclin levels. Since thromboxane may enhance the immune reactivity, stimulate platelet aggregation, cause vasoconstriction and smooth muscle cell proliferation, the increased levels of thromboxane have been postulated to be of pathogenetic importance for the development of chronic rejection. Prostacyclin, on the other hand, inhibits platelet aggregation and smooth muscle cell proliferation and also causes vasodilatation (*Sinzinger et al., 1987; Teraoka et al., 1987*).

The described factors are believed to be of importance for the development of chronic vascular rejection of transplanted organs. They are the basis for the strategies used in the treatment or prevention of chronic rejection. The pathophysiology is rather complex and no single factor could be pointed out to be more important than others. This is basically the reason why various approaches have been made with the aim to abrogate the chronic rejection process.

PREVENTION AND TREATMENT

There is very limited experience to date with therapeutic intervention, and most data emerge from experimental studies, while clinical studies have been limited and difficult to assess because of lack of accepted criteria to diagnose chronic rejection. Many of the studies are also uncontrolled and difficult to analyze because of concomitant drug therapies. Therapeutic strategies have largely been based on interference with metabolic disturbances considered to be of pathogenetic significance.

Since infiltrating inflammatory cells may participate in the development of intimal hyperplasia in chronic vascular rejection and since lymphocytes and monocytes may secrete destructive enzymes, cytokines and growth factors and participate in the formation of reactive oxygen radicals and enhance the *in situ* oxidation of the LDL fraction of lipoproteins and subsequently activate endothelial cells, it is fair to anticipate an important role of these cells in development of chronic vascular rejection. For this reason, it can also be assumed that chronic rejection is partly the result of *suboptimal immuno-suppression* (*Kuo and Monaco, 1992*). Therefore, the HLA match, the presence of one or more previous episodes of acute rejection, previous rejection of kidney transplants and presence of panel reactive antibodies may participate in development of chronic rejection. Furthermore, the commonly used immunosuppressive agents such as cyclosporine A and prednisolone may also have a pro-atherogenic potential. Cyclosporine A has been shown to have direct toxic effects on vascular endothelial and smooth muscle cells (*Zoja et al., 1986; Ferns et al., 1990*) and steroids may indirectly have a detrimental effect because of induction of hyperlipidemia and increased peripheral insulin resistance. On the other hand, some of the more recently developed and investigated drugs, like Rapamycin and RS-61443, seem to have advantageous effects on chronic rejection, based upon recent experimental results (*Morris et al., 1991*). Thus, optimizing the immunosuppressive treatment in organ transplantation may be an important first objective in the prevention of the development of chronic vascular rejection.

Prostaglandin modulators have been frequently tried, both clinically and experimentally, in the prevention and the treatment of chronic vascular rejection aiming at restoration of the thromboxane-prostacyclin imbalance present in renal chronic rejection. Continuous infusion of the synthetic prostacyclin analogue Iloprost® had a beneficial effect on the development of cardiac graft atherosclerosis in a rat model of long-graft survival, following limited cyclosporine treatment (*Fellström et al., 1991*). In preliminary reports on the use of prostacyclin analogues in renal transplant patients with an established chronic vascular rejection, beneficial effects were reported (*Teraoka et al., 1987b*). This study was made in a limited number of patients and with a limited observation period; it was an open, uncontrolled, pilot study with continuous intravenous infusion of epoprostenol for one week followed by a low dose of oral salicylate (100-150 mg daily) and dipyridamol (225 mg daily). In 22 patients with an established chronic renal allograft rejection, we saw a stabilization or improvement of function compared with pretreatment levels in 18/22 patients followed at least 9 months. These data suggest that treatment with prostacyclin analogues and cycloxygenase inhibitors may have some therapeutic potential, although controlled studies are still missing.

Thromboxane antagonists with anti-atherogenic potential, due to the properties of thromboxane to stimulate smooth muscle cell proliferation, platelet aggregation and cause vasoconstriction, have been tried in renal transplant patients with chronic rejection (*Teraoka et al., 1987a*). The thromboxane synthetase inhibitor OKY-046 was used and to date 20 patients were treated with OKY-046 and compared with 20 control cases. There was a significantly lower frequency of graft losses in the actively treated patients compared with the control group on a 12-month perspective (*Teraoka et al., 1992*).

Polyunsaturated fatty acids of the Omega-3 series cause a shift in the thromboxane and prostacyclin metabolites leading to a decreased platelet stickiness and aggregability, as well as a vasodilatation. Food supplementation with Omega-3 fatty acids causes a slight reduction in hypertriglyceridemia and also reduces the risk of development of atherosclerosis (*Weiner et al., 1986; Leaf and Weber, 1988*). Omega-3 fatty acids may be of particular interest in transplanted patients because they also enhance immunosuppressive properties of cyclosporine A (*Endres et al., 1989*). Furthermore, they decrease the synthesis of PDGF-β like peptides in endothelial cells. Renal transplant patients with a progressive decline in graft function due to chronic rejection were treated with MAXEPA supplementation, which resulted in a significant reduction in the rate of function deterioration compared with pretreatment levels (*Sweny et al., 1989; Sweny, 1992*). The treatment also caused a significant decrease in serum triglycerides and platelet aggregability. The authors are satisfied by the treatment results, and now they administer MAXEPA to all patients with this diagnosis.

The importance of hyperlipidemia for development of atherosclerosis is well established, and evidence has also emerged supporting the possible role of lipoproteins in the development of chronic vascular rejection. The use of cholesterol synthetase inhibitors in cardiac transplant patients reduced hypercholesterolemia without lowering HDL cholesterol (*Kobashigawa et al., 1990*). To date, there are no controlled studies showing an effect on the progression of chronic vascular rejection by cholesterol synthetase inhibitors or other lipid-lowering agents. As discussed above, there may be a rationale in using antioxidant agents in transplanted patients with chronic rejection. The antioxidant agent Probucol®, which has a weak cholesterol-lowering effect (*Steinberg, 1986*), is currently used in a clinical trial in renal transplant patients, but to date no results about the effect on chronic rejection have been reported.

Heparin and heparin-derivatives inhibit smooth muscle cell proliferation *in vitro* and *in vivo* (*Castello et al., 1982; Clowes and Clowes, 1989*) independent of their anticoagulant effect. Heparin may also attenuate the migration of smooth muscle cells into the intima and cause a shift in the matrix composition with a decrease in elastin and collagen and an increase in heparin sulphate and chondroitin sulphate (*Clowes and Clowes, 1989*). To what extent these effects are dependent on the potential of heparin to bind the b-FGF and PDGF is a hypothetical question. A protection of the vascular endothelium by heparin and heparin-derivatives seems to be present, however. The effect upon transplant intimal hyperplasia has been demonstrated experimentally with low molecular weight heparin in combination with cyclosporine (*Plissonnier et al., 1992*), and according to our own experience, there are heparin derivatives without anticoagulant effect which protect aorta grafts against development of intimal hyperplasia efficiently, but these experiments need to be extended further (unpublished results). To our knowledge, there are no clinical studies in transplant patients with heparin- like substances presented as yet.

The somatostatin analogue Angiopeptin is one of the most promising substances tested in the prevention and in the treatment of accelerated transplant atherosclerosis. Angiopeptin has all the common endocrine effects of a somatostatin, including reduced levels of growth hormone, insulin and glucagon and is considered to be a nontoxic and non-immunogenic peptide, which can be administered as a continuous intravenous infusion or as intermittent subcutaneous injections. Angiopeptin also inhibits cell proliferation in the rabbit aorta after balloon angioplasty (*Asotra et al., 1989; Conte et al., 1989*). Treatment with Angiopeptin also causes a 20-50% inhibition of myointimal proliferation in models of transplant atherosclerosis, including accelerated heart graft atherosclerosis in cholesterol fed and cyclosporine treated rabbits (*Foegh et al., 1989a; Foegh et al., 1989b*) and the rat (*Fellström et al., 1991*). In aorta transplants in the rat, Angiopeptin was also demonstrated to inhibit the intimal proliferation by 40% in syngeneic grafts and by 20% in allogeneic grafts, where the recipients were not treated with any immunosuppressive agent (*Fellström et al., 1992*). Today, Angiopeptin is on clinical trial in heart transplantation and in renal transplantation with the objective to prevent or to treat chronic vascular rejection or accelerated transplant atherosclerosis.

SUMMARY AND CONCLUSION

Chronic rejection is a major threat towards the long-term function and survival of heart and renal transplants. It is characterized by a proliferative remodelling of the graft vessels along with structural changes of the parenchyma and gradual deterioration of graft function. The pathogenesis is complex and multifactorial. Since grafts with chronic rejection are also subjected to a more or less intense invasion of immunoreactive cells, an important primary objective is to optimize the immunosuppressive treatment. There is no established means of prevention or treatment of chronic rejection, but pharmacological agents interfering with the prostaglandin metabolism have been tried most frequently. Preliminary results are also available from the use of polyunsaturated fatty acids and heparin derivatives. Based on experimental studies, the somatostatin analogue Angiopeptin seems very promising today. There will certainly be an increased interest in the use of lipid-reducing agents in the future as well as antioxidant agents acting against the effect of reactive oxygen radicals and oxidative modification of LDL lipoproteins.

REFERENCES

Asotra, S., Foegh, M., Conte, J.V., Cai, B.R. and Ramwell, P.W., 1989, Inhibition of ³H-thymidine in corporation by Angiopeptin in the aorta of rabbits after balloon angioplasty. *Transplant. Proc.*, 21:3695.

Cary, N.R.B., 1992, Diagnostic criteria of chronic rejection in transplanted hearts. *Transplant. Proc.* (in press).

Castellot, J.J. Jr, Favreau, L.V., Karnovsky, M.J. and Rosenberg, R.D., 1982, Inhibition of vascular smooth muscle cell growth by endothelial cell-derived heparin. Possible role of a platelet endoglycosidase. *J. Biol. Chem.* 257:11256.

Clowes, A.W. and Clowes, M.M., 1989, Inhibition of smooth muscle cell proliferation by heparin molecules. *Transplant. Proc.* 21:3700.

Conte, J.V., Foegh, M.L., Calcagno, D., Wallace, R.B. and Ramwell, P.W., 1989, Peptide inhibition of myointimal proliferation following angioplasty in rabbits. *Transplant. Proc.* 21:3686.

Cook, D.J. and Terasaki, P.I., 1989, Renal transplantation on the American Continent. *In: Brent, L. and Sells, R.A. (eds): Organ Transplantation. Current Clinical and Immunological Concepts.* Balliére Tindall, London, 195.

DiCorleto, P.E. and Bowen-Pope, D.F., 1983, Cultured endothelial cells produce a platelet-derived growth factor like protein. *Proc. Natl. Acad. Sci. U.S.A.* 80:1919.

Dimény, E., Fellström, B., Larsson, E., Tufveson, G. and Lithell, H., 1992, Hyperlipoproteinemia in renal transplant recipients - is it linked with chronic vascular rejection? *Transplant. Proc.* (in press).

Dunn, M., Crisp, S.J., Rose, M., Taylor, P.M. and Jacoub, M.H., 1992, Detection of antiendothelial antibodies by western blotting - positive correlation with coronary artery disease after cardiac transplantation. *Proc. NATO Adv. Study Inst.* (this issue).

Endres, S., Ghorbani, R., Kelly, V.E. et al., 1989, The effect of dietary supplementation with n-3 poly-unsaturated fatty acids on the synthesis of interleukin-1 and tumor necrosis factor by mononuclear cells. *N. Engl. J. Med.* 320:265.

Fellström, B., Dimény, E., Foegh, M.L., Larsson, E., Wanders, A. and Tufveson, G., 1991, Accelerated atherosclerosis in heart transplants in the rat simulating chronic vascular rejection: effects of prostacyclin and Angiopeptin. *Transplant. Proc.* 23:525.

Fellström, B., Dimény, E., Larsson, E., Claesson, K. and Tufveson, G., 1990, Rapidly proliferating arteriopathy in cyclosporin-induced permanently surviving rat cardiac allografts simulating chronic vascular rejection. *Clin. Exp. Immunol.* 80:288.

Fellström, B., Klareskog, L., Terracio, L., Larsson, E., Wahlberg, J., Tufveson, G., Rönnstrand, L., Heldin, C.H. and Rubin, K., 1989, Platelet derived growth factor receptors in the kidney - importance of up-regulated expression in renal inflammation. *Kidney Int.* 36:1099.

Fellström, B., Larsson, E., Claesson, K., Tufveson, G., Wahlberg, J. and Klareskog, L., 1988, Macro-phages and T-lymphocytes expressing HLA-D region encoded gene products in rejected renal transplants. *Transpl. Proc.* 20:372.

Fellström, B., Stafberg, C., Larsson, E., Aküyrek, M.L., Ren, Z., Wanders, A., Waltenberger, J., 1992, Angipeptin treatment of chronic vascular rejection in transplanted hearts and aortas in the rat. *Transplant. Proc.* (in press).

Ferns, G., Reidy, M. and Ross, R., 1990, Vascular effects of cyclosporine A *in vivo* and *in vitro. Am. J. Pathol.* 137:403.

Foegh, M.L., Khirabadi, B.S., Chambers, E., Amamoo, S. and Ramwell, P.W., 1989a, Inhibition of coronary artery transplant atherosclerosis in rabbits with Angiopeptin, an octapeptide. *Atherosclerosis* 78:229.

Foegh, M.L., Khirabadi, B.S., Chambers, E. and Ramwell, P.W., 1989b, Peptide inhibition of accelera-ted transplant atherosclerosis. *Transplant. Proc.* 21:3674.

Frostegård, J., Haegerstrand, A., Gidlund, M. and Nilsson, J., 1991, Biologically modified low density lipoprotein increases the adhesive properties of vascular endothelial cells. *Atherosclerosis* 90:1-19.

Frostegård, J., Nilsson, J., Haegerstrand, A., Hamsten, A., Wigzell, H. and Gidlund, M., 1990, Oxidized low density lipoprotein induces differentiation and adhesion of human monocytes and the mono-cytic cell line U937. *Proc. Natl. Acad. Sci. U.S.A.* 87:904.

Gao, S.Z., Schroeder, J.S., Alderman, E.L, et al., 1987, Clinical and laboratory correlates of acceler-ated coronary artery disease in the cardiac transplant patient. *Circulation* 76(Suppl.5):V56.

Hansson, G.K., Genk, Y., Holm, J. and Stemme, S. Lymphocyte adhesion and cellular immune reac-tions in chronic rejection and graft arteriosclerosis. *Transplant. Proc.* (in press).

Heldin, C.H. and Miyazono, K., 1992, Transforming growth factor-βS: Structural and functional properties. *Transplant. Proc.* (in press).

Kasiske, B.L., Kalil, R.S.N., Lee, H.S. and Rao, K.V., 1991, Histopathologic findings associated with a chronic progressive decline in renal allograft function. *Kidney Int.* 80:514.

Kobashigawa, J.A., Murphy, F.L., Stevenson, L.W. et al., 1990, Low-dose lovastatin safely lowers cholesterol after cardiac transplantation. *Circulation,* 82(suppl.4):IV-281.

Kuo, P. and Monaco, A.P., 1992, Chronic rejection as a result of suboptimal immunosuppression. *Transplant. Proc.* (in press).

Leaf, A. and Weber, P.C., 1988, Cardiovascular effects of n-3 fatty acids. *N. Engl. J. Med.* 318:549.

Lehr, H.A., Arfors, K.E., Hübner, C., Menger, M.D. and Messmer, K., 1992, Leukocyte/endothelium interaction as a target for anti-atherogenic strategies in allograft transplantation. *Transplant. Proc.* (in press).

Mahony, J.J. and Sheil, A.G.R., 1987, Long-term complications of cadaver renal transplantation. *Trans plant. Rev.* 47.

Maryniak, R.K., First, M.R. and Weiss, M.A., 1985, Transplant glomerulopathy: evolution of morphologically distinct changes. *Kidney Int.* 27:799.

Morris, R.E., Wang, J., Blum, J.R. et al., 1991, Immunosuppressive effects of the morpholinoethyl
ester of mycophenolic acid (RS-61443) in rat and nonhuman primate recipients of heart
allografts. *Transplant. Proc.* 23(Suppl.2):19.

Naparstek, Y., Cohen, I.R., Fuks, Z. and Vlodavsky, I., 1984, Activated Tlymphocytes produce a matrix
-degrading heparin sulphate endoglycosidase. *Nature* 310:241.

Nilsson, J., 1992, Lipid oxidation, vascular inflammation and coronary atherosclerosis. *Transplant. Proc.*
(in press).

Nilsson, J., Sjölund, M., Palmberg, L., Thyberg, J. and Heldin, C.-H., 1985, Arterial smooth muscle
cells in primary culture produce a platelet-derived growth factor-like protein. *Proc. Natl. Acad.
Sci. U.S.A.* 82:4418.

Patarroyo, M.P., Prieto, J., Rincon, J., Timonen, T., Lundberg, C., Lindbom, L., Asjö, B. and
Gahmberg, C., 1990, Leukocyte cell adhesion: a molecular process fundamental in leukocyte
physiology. *Immuno. Rev.* 114:68.

Paul, L.C., 1992, Chronic rejection of organ allografts: magnitude of the problem. *Transplant. Proc.* (in
press).

Paul, L.C., Häyry, P., Foegh, M., Dennis, M.J., Mihatsch, M.J., Larsson, E., Fellström, B., 1992, Diagnostic criteria of chronic rejection/accelerated graft atherosclerosis of heart and kidney transplants. Proposal from the Fourth Alexis Carrel Conference on Chronic Rejection and
Accelerated Arteriosclerosis in Transplanted Organs. *Transplant. Proc.* (in press).

Plissonnier, D., Amichot, G., Cecaqueux, J., Duriez, M. and Gentric, D., 1992, Synergy of a low
molecular weight heparin-like molecule and low doses of cyclosporin in preventing arterial
graft rejection in rats. (manuscript).

Regnström, J., Nilsson, J., Tornvall, P., Landou, C. and Hamsten, A., 1992, Susceptibility to low-density lipoprotein oxidation and coronary atherosclerosis in man. *Lancet* 339:1183.

Ross, R., 1986, The pathogenesis of atherosclerosis - an update. *N. Engl. J. Med.* 314:488.

Rubin, K., Tingström, A., Hansson, G.K., Larsson, E., Rönnstrand, L., Klareskog, L., Claesson-Welsh,
L., Heldin, C.-H., Fellström, B. and Terracio, L., 1988, Induction of B-type receptors for
platelet-derived growth factor in vascular inflammation: possible implication for development
of vascular proliferative lesions. *Lancet* i:1353.

Ruscetti, F.W. and Palladino, M.A., 1991, Transforming growth factor-β and the immune system. *Progress in Growth Factor Research* 3:159.

Shimokado, K., Raines, E.W., Madtes, D.K., Barrett, T.B., Benditt, E.P. and Ross, R., 1985, A significant part of macrophage-derived growth factor consists of at least two forms of PDGF. *Cell* 4
3:277.

Silver, B.J., Jaffer, F.E. and Abboud, H.E., 1989, Platelet-derived growth factor synthesis in mesangial
cells: induction by multiple peptide mitogens. *Proc. Natl. Acad. Sci. U.S.A.* 86:1056.

Sinzinger, H., Zidek, T. and Foegh, M.L., 1987, Platelet-derived growth factor and prostacyclin in kidney transplant rejection. *Transplant. Proc.* 19 (Suppl.l):l02.

Springer, T.A., 1990, Adhesion receptors of the immune system. *Nature* 346:425.

Steinberg, D., Parthasarathy, S., Carew, T.E., Kkoo, J.C. and Witztum, J.L., 1989, Beyond cholesterol,
modifications of low-density lipoprotein that increase its atherogenecity. *New Engl. J. Med.*
320:915.

Steinberg, D., 1986, Studies on the mechanism of action of probucol. *Am. J. Cardiol.* 57:16H.

Sweny, P., 1992, The use of dietary fish oils in renal allograft recipients with chronic vascular rejection.
Transplant. Proc. (in press).

Sweny, P., Wheeler, D.C., Lui, S.F. et al., 1989, Dietary fish oil supplements preserve renal function in
renal transplant recipients with chronic vascular rejection. *Nephrol. Dial. Transplant.* 4:1070.

Taylor, P.M., Rose, M. and Yacoub, M.H., 1992, Vascular adhesion molecules in heart transplantation.
Proc. NATO Adv. Study Inst. (this issue).

Teraoka, S., Takahashi, K. and Toma, H., 1987a, Application of prostacyclin analogue and
thromboxane synt-hetase inhibitor to chronic vascular rejection after kidney transplantation.
Transplant. Proc. 19:36-64.

Teraoka, S., Takahashi, K., Toma, H., Ota, K., 1992, Controlled prospective study of treatment for
chronic rejection after kidney transplantation by thromboxane synthetase inhibitor. *Transplant.
Proc.* (in press).

Teraoka, S., Takahashi, Y., Toma, H. et al., 1987b, New approach to management of chronic vascular
rejection with prostacyclin analogue after kidney transplantation. *Transplant. Proc.* 19:215.

Thorogood, J., Van Houwelingen, H.C., Van Rood, J.J., Zantvoort, F.A., Schreuder, G.M.Th. and Persijn, G.G., 1992, Long-term results of kidney transplantation in Eurotransplant. In: Paul, L.C. and Solez, K. (eds): Organ Transplantation: Long Term Results. Marcel Dekker, Inc., New York (in press).

Waltenberger, J., Wanders, A., Fellström, B., Larsson, E., Miyazono, K., Funa, K. and Heldin, C.-H., 1992, Expression of transforming growth factor-β in heart transplants in the rat. *Transplant. Proc.* (in press).

Wanders, A., Akyürek, M.L., Larsson, E., Zhiping, R., Stafberg, C., Waltenberger, J., Funa, K. and Fellström, B., 1992, Effect of graft ischemia time on the development of experimental transplant arteriosclerosis. *Transplant. Proc.* (in press).

Weiner, B.H., Ockene, I.S., Levine, P.H. et al., 1986, Inhibition of atherosclerosis by cod-liver oil in a hyperlipidemic swine model. *N. Engl. J. Med.* 315:841.

Zoja, C., Furci, L, Ghilardi, F., Zilio, P., Benigni, A. and Remuzzi, A., 1986, Cyclosporine-induced endothelial cell injury. *Lab. Invert.* 55:455.

VI. EPILOGUE

1992 NATO ASI CONFERENCE ON "VASCULAR ENDOTHELIUM:

PHYSIOLOGICAL BASIS OF CLINICAL PROBLEMS II"

John Gordon

Director of Research
British Bio-Technology Limited
Watlington Road, Cowley
Oxford. OX4 5LY
United Kingdom

The invitation to chair the final Panel Discussion at the end of a prestigious meeting like this was difficult to refuse, especially when the invitation was extended by someone as persuasive as John Catravas (in a similar context, Judah Folkman once remarked that the gene for effective flattery is present in only a few gifted individuals, but the receptor is widely distributed). Having accepted the invitation, the Chairman of the final session then basks in the rosy glow emanating from his well-massaged ego until the time comes (usually the evening before the final session) to review the full enormity of the task before him.

And to write a summary of the final session is an even greater challenge ("two pages on highlights, consensus and future directions should be adequate" said the good Dr. Catravas). What kind of summary is appropriate? A brief recapitulation of each presentation is not necessary, as they are all in this book: summarizing selected presentations would introduce invidious distinctions (there is always the temptation to plug one's own presentation - and, like Oscar Wilde, I can resist everything except temptation - but considerations of good taste and scientific integrity should prevent any rapporteur from drawing undue attention to his own presentation). Yet there are clearly highlights to be emphasized, and several elements of consensus emerged from our final discussion, notably with respect to the future directions that meetings in this series might take. Therefore, in the following two pages or so, I have attempted to capture something of this meeting which I hope will prove useful both for those who were there and for those who have to rely on this volume for their information.

The first highlight of the meeting was the evident care that had gone into its planning, and this has an important message for organizers of future meetings. Careful planning showed clearly in the structure of the program, which combined themes of different pathologies and the mediators responsible for them, and also introduced relevant therapeutic approaches, both current and potential. The diversity of the Faculty was another important element in the planning of the meeting: there were examples of the young, the well-known specialists, and those who could survey the whole field, thus representing the first three of the categories of scientific speakers listed by Victor Weisskopf, the ex-director-General of CERN (Weisskopf also listed two even more distinguished categories - those who deliver memorial speeches about deceased colleagues and those who only give after-dinner speeches - but neither of these was evident at our meeting). Finally, the venue represented another triumph of planning, providing excellent facilities in splendid surroundings. It has been rumored that the reason the three meetings in this series have all been held in relatively remote

Vascular Endothelium, Edited by J.D. Catravas *et al.*
Plenum Press, New York, 1993

(and different) parts of Greece is because these locations are all far enough away from the director's homebase that the conference hotels there are prepared to accept his booking because they don't know his reputation. However, those of us who know John Catravas find that allegation difficult to believe.

The second obvious highlight was the efficiency of the organization at the meeting itself. This was manifest in smooth administration with a minimum of staff. There was absolutely no evidence of contamination by the element known as "administration", which scientists at the Harwell Atomic Energy Research Station recently described as the heaviest known. It apparently possesses one neutron, 8 assistant neurons, 64 vice-neurons and 256 assistant vice-neurons and the whole structure is kept in place by particles call morons. "Administratium" is inert, but it impedes every reaction with which it comes into contact. The organization of this ASI could not have been more different. In addition, the staff at the Conference hotel represented Greek hospitality at its best: I observed a sign that said "Visitors are expected to complain at the office between the hours of 9 and 10 a.m. daily" but I doubt that this facility was ever needed. (Another hotel displayed a notice "Flattening of underwear with pleasure is the job of the chambermaid" but this service was apparently not offered at the Paradise Hotel in Rhodes).

One of the most important highlights of the meeting was the formal scientific program itself. This covered a defined spectrum of vascular and inflammatory pathologies: it was deliberately not comprehensive, but individual, important themes were selected that allowed the logical development from basic science through to the clinic to be followed clearly. The main pathologies covered were: atherosclerosis and coronary heart disease; vasospasm; transplant rejection; and certain aspects of neoplasia. Other pathologies, including shock, diabetes and malaria, were also touched on, although with less emphasis. One theme running through many of the presentations was that a single biochemical mediator could play an important role in several different pathological conditions. Mediators that figured prominently in the meeting were reactive oxygen species (including NO); cytokines; and cell surface adhesive molecules. Discussions on therapy (actual or potential) included chemical drugs, monoclonal antibodies, surgery and gene transfer. The common theme throughout all the presentations was, of course, the central role of the vasculature - because the vascular endothelium, being the interface between the blood and the rest of the body, is therefore the location at which many pathologies have their genesis.

The discussions, both formal and informal, were another highlight of the meeting - they were generally well-informed, good-natured and often forthright. Informal discussions outside the scientific sessions are a particularly valuable source of information: I was extolling the virtues of a short-acting tranquillizer which allowed me to rest during a long flight and give a good presentation immediately after it, when one of my learned colleagues pointed out that a well-known side-effect of that particular drug was that you thought you were performing brilliantly when you were actually talking nonsense. It was worth going to Rhodes for that incisive comment alone.

What lessons can be learned regarding the directions to be followed at a subsequent ASI? I would propose the following. First, that the program should again focus on a clearly defined track from basic science through to the clinic, with a selection of relevant pathologies, biochemical mediators and approaches to therapy. Secondly, any temptation to overload the program with complexity should be avoided. Over complexity can arise if presentations are included for their individual brilliance, rather than their relevance; if they form a disparate collection, and not a unified whole, much of the value is lost. John Harlan's presentation included Larson's cartoon in which a schoolboy asks to be excused because his brain is full; the risk of experiencing that sensation is something we face as we attempt to deal with "information overload" in our rapidly-expanding area of scientific research, and a commonality of themes at a meeting can help with the problem.

Specific recommendations that emerged from our final discussion included pathologies that might be included in the next program, such as chronic inflammatory conditions (e.g. the arthritides); mechanisms of tumor growth and invasion (e.g. the metastatic process); metabolic vascular pathologies (e.g. diabetes); and the aging process itself. For presentations that concentrated on biochemical mediators, Bruce Freeman made the point that "targeted molecule reactions" are often poorly understood, and he suggested a series of provocative questions that could profitably be addressed: Who is the bad guy? Where did he come from? What did he do? How do we eliminate him? Is he all bad? It was specifically emphasized that, where possible, we should explore the effects of mediators *in vivo*, not just *in vitro* (though the latter is usually much easier, and we tend to measure what we can, rather than what we should). Also, we should be alert to the effects of flow on the reactions being discussed. To help link the studies on mediators with the clinical presentations, it was suggested that "mechanism experts" should be asked to present their work in a pathological context - and not merely in relation to one single disease.

Finally, it was frequently emphasized that opportunities for informal contacts during the meeting should be a high priority. These are where much information is exchanged, collaborations are initiated and differences can be resolved. The opportunities are many and varied - I have discussed esoteric research topics while bouncing along on a rented moped down Greek tracks that were more like dried riverbeds (yes, it did feel like moving from one ex-stream to another). Other delegates establish contacts in the bar or the disco (the latter is not advised for the more mature members of the Faculty - it can easily become the slipped disco). The precise location does not matter, but what does matter is the people - human beings have been described as the most efficient and cost-effective non-linear programming machines that can be mass-produced using unskilled labor, and they are also the key element in any successful scientific meeting, providing the program does justice to the subject. The quality of the participants at this ASI contributed greatly to its success, and the organizers and sponsors of the meeting have every reason to be pleased with the outcome. The prospects for the next one are excellent.

VII. ABSTRACTS OF ORAL AND POSTER PRESENTATIONS

ANTIBODIES TO ELASTIN-DERIVED PEPTIDES IN SERA OF

HEALTHY SUBJECTS AND ATHEROSCLEROTIC PATIENTS

S. Baydanoff and G. Nicoloff

Department of Biology, Medical University
5800 Pleven
Bulgaria

Healthy subjects of different ages and a group of atherosclerotic patients were tested for the presence of antielastin antibodies of different immunoglobulin classes (IgG, IgM, IgA and IgD) by the enzyme-linked immunosorbent assay (ELISA). All tested sera showed detectable levels of antielastin antibodies of the four immunoglobulin classes. In healthy subjects, antielastin IgG and IgM showed a relatively high level in the sera of children, growing even higher in the sera of 18-20 year old subjects. Then their levels were stabilized in the 30-60 year old for IgG and the 30-50 year old for IgM and gradually decreased thereafter. The antielastin IgA showed insignificant changes up to the age of 40 and then their level gradually increased. The antielastin IgD showed statistically insignificant changes with age and only a tendency to decrease after the age of 60. The atherosclerotic patients showed statistically significant decrease in antielastin IgG, IgM and IgD and significant increase in the antielastin IgA in comparison with controls of the same age.

Further investigation will be necessary to clarify the correlation between the level of antielastin IgA and elastin activity in the vascular wall.

153

BRADYKININ STIMULATES L-ARGININE TRANSPORT AND NITRIC

OXIDE RELEASE IN VASCULAR ENDOTHELIAL CELLS

Richard G. Bogle, S.B. Coade, *S. Moncada, J.D. Pearson and G.E. Mann

Vascular Biology Research Center
Biomedical Sciences Division
Kings College
Campden Hill Road
London
United Kingdom

*Wellcome Research Laboratories
Beckenham, Kent
United Kingdom

Vascular endothelial cells synthesize nitric oxide (NO) from the terminal guanidine-nitrogen atom of the cationic amino acid L-arginine (*Palmer, R. et al., 1988*). Although basal NO synthesis is not limited by arginine availability, under conditions of stimulated NO production or following substrate deprivation exogenous L-arginine appears to be rate-limiting (*Palmer, R. et al., 1988*). We have now investigated whether bradykinin, an agonist which is known to release NO release, modulates the endothelial cell L-arginine transporter.

Microcarrier cultures of 24 h arginine-deprived porcine aortic endothelial cells were perfused continuously with a HEPES-buffered Krebs solution containing L-[^{3}H]arginine and D-[^{14}C]mannitol (extracellular reference). Fractions of the column effluent were collected to measure either uptake of L-[^{3}H]arginine (*Mann, G.E. et al.,* 1989) or NO production. As described previously (*Bogle, R.G. et al., in press; Ishii, K. et al., 1991*) NO release was measured by monitoring cGMP production in LLC-PK$_1$ cells exposed for 1.5 min to the endothelial cell column effluent. In control experiments LLC-PK$_1$, cells did not show any significant change in cGMP levels following incubation with bradykinin (100nM, 1.5 min), however, levels were markedly elevated in LLC-PK$_1$ cells exposed to sodium nitroprusside (10 nM, 1 min), allowing use these cells as suitable target cells for the measurement of NO.

Transport of L-arginine was stimulated 2 ± 0.4 fold following a 2 min infusion of 100 nM bradykinin through the endothelial cell column. In parallel experiments bradykinin also stimulated release of NO (2 ± 0.5 fold) and prostacyclin (106 ± 32 fold). Nitro-L-arginine, a selective inhibitor NO synthase, had a negligible effect on both basal and bradykinin-stimulated L-arginine transport whereas agonist-induced NO release was virtually abolished (1 ± 0.2 fold). Removal of extracellular calcium shortened the duration of the bradykinin-induced elevation in arginine transport and inhibited NO release. Furthermore, the effects of bradykinin on transport and NO release were ablated during perfusion with Krebs containing 70 mM K$^+$, suggesting that the increased entry of L-arginine may be secondary to a membrane hyperpolarization induced by bradykinin.

154

CYTOKINE MODULATION OF ADULT AND NEONATAL LEUKOCYTE ADHERENCE TO VASCULAR ENDOTHELIUM

*,**Anne Burke-Gaffney, *T.J. Flood and **A.K. Keenan

*Children's Research Center
Our Lady's Hospital for Sick Children
Crumlin, Dublin 12
Ireland

**Department of Pharmacology
University College
Dublin
Ireland

Neutrophil (PMN) adhesion to cytokine activated endothelial cells (EC) is an early and requisite event in acute inflammation. Adhesion of lymphocytes to both activated and normal endothelium is a crucial step in immune surveillance. Immaturity of neonatal PMN and T lymphocytes may dramatically reduce their ability to adhere to endothelium at sites of inflammation. This may contribute to the increased susceptibility of the neonate to infection.

We have used an *in vitro* adhesion assay, based on the uptake of the vital dye Rose Bengal by both leukocytes and human umbilical vein endothelial cells (HUVEC), to compare the adhesion of adult and neonatal (cord blood) PMN and T cells to HUVEC. In the case of PMN, quantification of the adhesion molecules, leukocyte endothelial cell adhesion molecule (LECAM-1) and the integrin CD 11b/CD18 (MAC-1) allowed a comparison between these levels and the degree of adhesion.

HUVEC were seeded at a density of 4×10^4 cells/well on to gelatinized 96 well plates. PMN were isolated by discontinuous Ficoll/Hypaque gradient centrifugation and T cells were purified from peripheral blood mononuclear cells (PBMC) by passage over nylon wool columns. Adhesion was determined by incubation of PMN (30 min, 5×10^5 cells/well) or T cells (60 min, 4×10^5 cells/well) with unactivated or interleukin (IL)-1β activated HUVEC, followed by removal of non-adherent cells by washing and staining of HUVEC and adherent leukocytes with Rose Bengal (0.25%, 5 min). The color was developed on incubation with PBS: ethanol (1:1) for 60 min. Absorbance was read at 570nm. Results were expressed either as absorbance values or as % or fold increase in adherence (i.e. increase in OD at 570nm). LECAM-1 and MAC-1 expression were determined by flow cytometry using Leu-8 monoclonal antibody (mAb) and anti CR3 mAb respectively. Results expressed as mean channel no. Results are presented as mean ± SD of n determinations. Statistical analysis was performed using the Mann-Whitney-Wilcoxon test. Differences were deemed significant if $p < 0.05$.

A linear relationship was initially established between increasing cell number (HUVEC, PMN) and absorbance at 570nm on uptake of Rose Bengal. Linear regression analysis of Leu-8 expression on unactivated PMN vs. increase in adherence gave a correlation value of 0.717 ($p < 0.01$). Increase in adherence of unactivated adult

PMN to cytoline activated HUVEC (IL-1β, 5 U/ml, 4h)(73.7 ± 17.6%,n=7) was significantly greater (p<0.005) than neonatal PMN adherence (37.2 ± 18.3%, n=7). This is paralleled by a similar significant difference (p<0.005) in Leu-8 expression on the same cell populations. (Adult = 108.3 ± 43.9, neonate (28.2 ± 9.3).

Adherence of cytokine activated (Granulocyte macrophage colony stimulating factor (GMCSF) 50U/ml, 30 min) PMN to IL-1β activated HUVEC was significantly greater (43.6 ± 14.5%, n=7, p<0.05) than neonatal PMN adherence (28.5 ± 15.5%, n=7). This was paralleled to similar significant (p<0.005) differences in % increase in CR3. (Adult = 64.7 ± 15.8%, neonate = 10.3 ± 11.8%). Finally, preliminary data indicated that while unstimulated adult and neonatal T cells bind IL-1β(6h,20U/ml) activated HUVEC to a similar extent, it would appear that neonatal T cells (stimulated with phorbal dibutyrate (PDBU), 1μM, 24h) bind less effectively than stimulated adult T cells to unstimulated HUVEC.

In conclusion, therefore, it would appear that unstimulated and cytokine stimulated neonatal PMN bind less efficiently to cytokine activated HUVEC than do adult PMN, results paralleled by lower levels of expression (LECAM-1) or smaller increases (MAC-1) in cell surface adhesion molecules. Similar degrees of binding of both adult and neonatal T cells to cytokine stimulated HUVEC may be accounted for by similarities between cytokine activated HUVEC and high endothelial venules of the lymph nodes. Lesser binding of stimulated neonatal T cells compared to adult, to unactivated HUVEC, may reflect the inability of naive T cells to respond to mitogenic stimuli.

This data supports the hypothesis that immature or inadequate host defence mechanisms may be causally related to susceptibility of the human neonate to infections. *(This work was funded by the Children's Research Center.)*

INTER-COMMUNICATION BETWEEN BRAIN CAPILLARY

ENDOTHELIAL CELLS AND ASTROCYTES: UPREGULATION

OF THE LDL RECEPTOR AT THE BLOOD-BRAINED BARRIER

Roméo Cecchelli, Bénédicte Dehouck, Marie Pierre Dehouck
and Jean Charles Fruchart

SERLIA-INSERM U325
Pasteur Institute
59019 Lille Cédex
France

The presence of lipoproteins, apolipoproteins and their receptors in the brain could provide a system for cholesterol homeostasis, as they do in other tissues. Our previous work demonstrating *in vivo* the occurrence of a low density lipoprotein (LDL) receptor on the BBB has suggested that this cerebralendothelial receptor plays a role in lipid transport across the BBB (*Meresse, S. et al., 1989*). To examine further the function of this receptor, we have established an *in vivo* model, that closely mimics the in vivo situation, by co-culturing brain capillary endothelial cells on one side of a filter and astrocytes on the other (*Dehouck, M.P. et al, 1990*). When cultured separately, the brain capillary endothelial cells express a LDL receptor with the same apparent molecular weight as *in vivo*. When the astrocytes and the brain capillary endothelial were co-cultured, the number of LDL receptors located at the luminal side increased by ~3 fold.

To ascertain whether co-culturing for 8 days influences the expression of the LDL receptor on brain capillary endothelial cells, astrocytes were cultured in lipoprotein deficient serum (LPDS) to decrease the lipid concentration of the cells. When the two cell types were again co-cultured, we noted that after 3 hours LDL binding increased by 600%. These results demonstrate that the lipid composition of astrocytes modulates the expression of the LDL receptor at the blood-brain barrier interface.

Substituting astrocytes with smooth muscle cells or, on the other hand, substituting brain endothelium with endothelium from aorta in the co-culture did not enhance the expression of the LDL at the luminal side of endothelial cells. Furthermore, we have shown that this modulation occurs only when the two cell types were in "long-term" co-culture, demonstrating that astrocytes and brain endothelium interact specifically *in vitro* during the co-culture. Since these effects in co-cultures were mediated through a filter that prevents direct cell-cell contact, we conclude that soluble factors derived from astrocytes are responsible for the induction of the LDL receptor. This hypothesis was confirmed by the fact that conditioned medium from astrocytes in long term co-culture in LPDS increased the expression of the LDL receptor at the luminal side of the *in vitro* BBB.

REFERENCES

Méresse, S., Delbart, C., Fruchart, J.C. and Cecchelli, R., 1989, LDL receptor on endothelium of brain capillaries. *J. Neurochem.* 53:340-345.

Dehouck, M.P., Méresse, S., Delorme, P., Fruchart, J.C. and Cecchelli, R., 1990, An easier reproducible and mass-producing method to study the blood-brain barrier. *J. Neurochem.* 54:1798-1801.

PMA-ACTIVATED NEUTROPHIL-MEDIATED ENDOTHELIAL ANGIOTENSIN CONVERTING ECTOENZYME DYSFUNCTION

Xilin Chen and John D. Catravas

Department of Pharmacology and Toxicology
Medical College of Georgia
Augusta, GA 30912
U.S.A.

We studied the effects and mechanisms of activated rabbit peritoneal neutrophils (PMN) on endothelial-bound angiotensin converting enzyme (ACE) activity in cultured bovine pulmonary arterial endothelial cells (EC), using [^{3}H]-Benzoyl-Phe-Ala-Pro as the ACE substrate under first order reaction condition. Phorbol myristate acetate (PMA) or PMN alone had no effects on ACE activity. When PMN were co-incubated (activated) with PMA (10 ng/ml) for 4 hours in Earle's salt solution, endothelial ACE activity was decreased 87%. No EC cytotoxicity was noticed at this stage as determined by ^{51}Cr release from pre-labelled EC. PMN-mediated ACE dysfunction was inhibited by catalase (2,000 U/ml), but not by superoxide dismutase (300 U/ml). PMN-mediated ACE dysfunction was also inhibited by the hydroxyl radical scavenger dimethyl thiourea (5 mM) but not mannitol (5 mM), which does not cross the cell membrane. Pre-treatment of EC with the iron chelator deferoxamine mesylate (1 to 10 mM) for 4 h attenuated the PMN-mediated ACE dysfunction, while co-incubation deferoxamine did not affect the activated PMN-mediated ACE dysfunction. The thiol reducing agent, 2-mercaptoethanol (0.1 mM) also prevented PMN-mediated ACE dysfunction. The myeloperoxidase inhibitor, cyanide (5 mM), but not azide (1 to 50 mM), also inhibited the activated PMN-induced EC ACE dysfunction. Treatment with the proteinase inhibitor phenylmethylsulfonyl fluoride or with human α-antitrypsin did not affect the action of activated PMN. NO synthase inhibitor, N^w-nitro-L-arginine (0.1 mM), also had no effects on PMN-mediated ACE dysfunction. These results suggest that PMN-mediated ACE dysfunction may be due to the production of hydrogen peroxide by PMN and its conversion into hydroxyl radicals.

A NOVEL METHOD FOR CULTURING VASCULAR ENDOTHELIAL CELLS AND STUDYING ADHESION OF BLOOD CELLS UNDER CONDITIONS OF FLOW.

Brian M. Cooke, *S. Usami, I. Perry and G.B. Nash

Department of Hematology
The Medical School
Birmingham B15 2TT
United Kingdom

*Institute of Biomedical Sciences
Academia Sinica
Taiwan

Adhesion of blood cells to vascular endothelial cells may contribute to microcirculatory obstruction and ischemia in a variety of pathological conditions. Studies of such interactions *in vitro* are best carried out using assays that mimic flow *in vivo*. It is also widely accepted that hemodynamic flow forces influence the structure and regulate the function of the endothelial cells themselves.

We have therefore developed a simplified technique to culture human umbilical vein endothelial cells, (HUVEC), under shear-flow conditions, in pre-fabricated, glass, microcapillary tubes, ("microslides"), with well-defined rectangular cross section and good optical quality. The endothelialized microslides have been incorporated into a controlled flow system for quantitative video-microscopic analysis of the adhesion of blood cells to endothelial cells.

Microslides were pre-treated with 3-aminopropyltriethoxysilane and then loaded with a suspension of endothelial cells. After the cells had settled and attached to the substrate, the microslides were inserted into a flow-based culture system. Culture medium was drawn through them at intervals, or continuously, until confluency was reached (approximately 24 hours).

For adhesion assays, the endothelialized microslides were attached to microscope slides, and suspensions of blood cells were drawn through at desired wall shear stresses (0.02-0.5 Pa). Using this methodology, we have demonstrated inhibition of adhesion of neutrophils and homozygous sickle cells to HUVEC by pharmacological agents and quantitated the strength of the adhesive interaction between red blood cells harboring the malarial parasite, *Plasmodium falciparum*, and endothelial cells.

A transportable form of the adhesion assay, using a hydrostatic flow system and formalin-fixed HUVEC enabled us to carry out a field study in West Africa. This demonstrated, for the first time, that red cells parasitized by laboratory and wild strains of *P. falciparum* are able to adhere to HUVEC at physiologically relevant wall shear stresses.

The technique, is therefore well suited for analyzing adhesive phenomena and for screening putative inhibitors of adhesion which may have therapeutic potential.

DRUG TRANSFER ACROSS THE BLOOD-BRAIN BARRIER: CORRELATION

BETWEEN *IN VITRO* AND *IN VIVO* MODELS

M.P. Dehouck, B. Dehouck, J.C. Fruchart, R. Cecchelli
and *P. Jolliet-Riant, F. Bree, J.P. Tillement

U325 Institut Pasteur de Lille
59019 Lille Cédex
France

*Laboratoire Hospitalo-Universitaire de Pharmacologie
UFR de Médecine de Paris XII
8 rue du Général Sarrail
94010 Créteil
France

To provide an *in vitro* system for studying brain capillary functions, we have developed a process of co-culture that closely mimics the *in vivo* situation, by culturing brain capillary endothelial cells on one side of a filter and astrocytes on the other. In these conditions, endothelial cells retain all the endothelial cell markers and the characteristics of the blood-brain barrier. To validate our *in vitro* blood-brain barrier model, we have compared the transport of ten different compounds across the *in vitro* model vs. transport across the blood-brain barrier *in vivo*. *In vivo* brain extraction (Eo) are measured according to the method of Oldendorf. The *in vivo* and *in vitro* Eo values show a strong correlation as indicated by the Spearman's co-efficient ($r = 0.88$, $p < 0.01$). The *in vitro* blood-brain barrier permeability for glucose and leucine i.e. compounds that traverse the blood-brain barrier via carrier mediation is in the same range as the blood-brain barrier permeability *in vivo* indicating that in our model the glucose and leucine transporters are always present. The relative ease with which such co-cultures can be produced in large quantities would facilitate the investigations in delivery of nutriments across the blood-brain barrier, and the screening of new centrally acting drugs.

BOVINE NEUTROPHILS IMPAIR ENDOTHELIUM-DEPENDENT RESPONSES TO ACETYLCHOLINE AND NORADRENALINE IN ISOLATED MESENTERIC ARTERIES

Sjef J. De Kimpe, Dicky Van Heuven-Nolsen and Frans P. Nijkamp

Department of Pharmacology
Faculty of Pharmacy
University of Utrecht
3508 TB, Utrecht
The Netherlands

Objective: Several pathological states are associated with neutrophil accumulation and activation. The interaction of neutrophils with blood vessels may be important in the regulation of vascular tone. In the present study, we examined the effect of polymorphonuclear leukocytes (PMN) on the endothelium-dependent relaxation to acetylcholine and on the vasoconstriction evoked by noradrenaline in the bovine isolated mesenteric artery.

Methods: Therefore, the mesenteric arterial rings were incubated with bovine PMN (2.5×10^6 ml^{-1}) in an organ bath. The force was measured isometrically. After a 20 min incubation period the arteries were precontracted with PGF_{2a} and relaxation to acetylcholine was determined.

Results: The vasodilatation to acetylcholine was decreased in the presence of PMN (Emax ± sem: 75.8 ± 2.5 % for control vs. 44.9 ± 7.4% for PMN, <0.001). In contrast, the endothelium-independent vascular relaxation to nitroprusside was not diminished by PMN. Furthermore, the sensitivity to L-noradrenaline was augmented in the presence of PMN (D2 ± sem: 6.46 ± 0.06 for control and 6.86 p m 0.07 for PMN, p < 0.001). The maximal contraction was not influenced by PMN. This effect was endothelium dependent. The impaired relaxation to acetylcholine and the increased sensitivity to noradrenaline in isolated mesenteric arteries induced by PMN was completely inhibited by superoxide dismutase, but not by other oxygen-radical scavengers or inhibitors of arachidonic acid metabolism. This PMN-induced effect was reversible, since the relaxation to acetylcholine was not influenced, after the PMN had been washed out of the organ bath before the precontraction.

Conclusion: We argue that PMN-derived superoxide anions contribute to a decrease in relaxation to acetylcholine and an increase in noradrenaline sensitivity by interference with endothelium-derived relaxing factors. *(This study was subsidized by the Netherlands Heart Foundation, NHS 88.248).*

DECREASED RELAXATION TO BRADYKININ AFTER INTERFERON-γ IN

BOVINE ISOLATED MESENTERIC ARTERIES

Sjef J. De Kimpe, Wanda C.M. Tielemans
Dickky van Heuven-Nolsen and Frans P. Nijkamp

Department of Pharmacology
Faculty of Pharmacy
University of Utrecht
3508 TB, Utrecht
The Netherlands

Objective: Cytokines are important in the regulation of the immune function. It is now apparent that tissue cells other than leukocytes are important ratgets of cytokines. In the present study, we investigated the influence of IFN-γ on the endothelium-dependent relaxation to bradykinin.

Methods: Arterial rings were isolated from the bovine mesenterium and incubated with bovine recombinant IFN-γ for 20 hr at 370C. subsequently, relaxation was measured in an organ bath after U46619-precontraction.

Results: The endothelium-dependent relaxation to bradykinin in mesenteric arteries was decreased in a concentration dependent manner by incubation with IFN-γ (1-1000 U/ml). Interestingly, the rings showed contraction instead of relaxation at the highest bradykinin concentrations ($10-8$ - $3x10-7$ M) after IFN-γ (Emax + sem expressed as percentage of the vascular tone induced by U46619: +7 + 11% for 100 U/ml IFN-γ vs. -84 + 4% for control). The effect was not found after a short 2 hr incubation period. Furthermore, simultaneous incubation of IFN-γ with cycloheximide completely prevented the decrease in relaxation. A small inhibition was found with verapamil, but L-NMMA largely inhibited the effect of IFN-γ. The endothelium-independent vascular relaxation to nitroprusside was also diminished by 20 hr incubation with IFN-γ (Emax | sem: =64 + 6% for control vs. -37 + 10% for 100 U/ml IFN-γ).

Conclusion: We conclude that IFN-γ induces a decrease in vasorelaxation to bradykinin, via the synthesis of a yet unknown factor. This effect is possibly mediated by interference with the action of nitric oxide. (*This study was subsidized by the Netherlands Heart Foundation*, 88.248).

ULTRASTRUCTURAL OBSERVATIONS ON ANGIOGENETIC CELL FORMATION AND DIFFERENTIATION TOWARDS VASCULAR ENDOTHELIUM IN HUMAN PLACENTAL VILLI

Ramazan Demir, *Peter Kaufmann, **Türkan Erbengi

Department of Histological Embryology
Faculty of Medicine
Akdeniz University, Antalya

*Marmara University
Istanbul
Turkey

**Department of Anatomy
RWTH, Aachen
Germany

Human placental villi undergo extensive morphogenesis during the earliest stages of development, and the predominant cell type of chorionic villi core, with a well-developed basement membrane separating the trophoblast from their stroma, were indifferent mesenchymal cell and fibroblast-like cell types with well-developed GER cisternae and Golgi complex (*Demir et al., 1987*). The basic cell types of villous core are of indifferent mesenchymal cell variety. In the early periods of pregnancy, they show a differentiation towards whole variety of villous stromal cells, some of them constituting the early angioblastic masses (*Demir and Erbengi, 1987; Demir et al., 1989*) and forming the blood vessels. These new cells are fibroblasts, reticulum cells, pericytes, endothelial cells, Hofbauer cells and embryonal plasma cells. They construct all stromal architecture.

The aim of this study is to investigate the formation and differentiation of angiogenetic cells to human placental villi. The development and the differentiation of the chorionic villi core have been investigated in human placental villi of the first, second and third trimesters of pregnancy. Placental villi of 12 exactly defined early human specimens ranging from day 21 post conception (p.c.) to day 42 p.c. and from an additional 6 specimens from about 7 to 40 weeks menstrual age have been analyzed ultrastructurally with regard to fetal vasculogenesis and angiogenesis.

The most primitive forms of vascular tube or capillary have been distinguished at 4th week of gestation. Contrary to the conclusion of Herring (1935) the primitive vascular tubes and their cellular elements were derived from extraembryonal mesoderm not cytotrophoblast, and they were composed of two or more angiogenetic cells attached to each other by tight junctions or desmosome-like complexes, not syncytium. The following results were obtained: The first cells differentiating at day 21 p.c., probably originating from mesenchymal precursors, are macrophage-like cells. Capillary formation takes place by the aggregation of haemangioblastic cells which are attached to each other by intercellular junctions. The lumen is formed by the dehiscence of the intercellular clefts.

Most of capillaries were surrounded by the pericytes and their cytoplasmic processes were closely attached to the outer surfaces of endothelial cells and no basement membrane was observed around the fetal capillaries.

The hunchbacks of pericytes were inserted between adjacent endothelial cells so that the pericytes participate in the formation of capillary walls. The vascular development was completed at 20 th or 22 th week of gestation. In this last period of gestation, fetal villous angiogenesis takes place by the proliferation of the existing endothelium and pericytes rather than via haemangloblastic cells.

In conclusion, (1) haemangiogenetic cells derive form indifferent mesenchymal cells, which transform toward haemangloblasts, (2) primitive vascular tubes form by haemangloblbiants and presumptive pericytes, (3) primitive capillary lumens appear at extracellular compartments surrounded by presumptive endothelium, (4) a capillary basal lamina cannot be detected in earlier periods, (5) in the last period of gestation fetal villi angiogenesis takes place by the proliferation of the existing endothelium and pericytes rather than via haemangioblastic cells.

REFERENCES

Demir, R., Erbengl, T., Kaya, M., 1987, Insan plascutasinda koryonik villustarin damarlanmasiüzerinde ultrastrüktürel bir Çaeiçma. *Türkiye Klinlkieri Tip Billmieri Arasurma Dergisi* 5(5):431-444.

Demir, R., and Erbengl, T., 1987, Vasculogenesis of human placental villi in early pregnancy. *Acta Anatomica* 130:25.

Demir, R., Kaufmann, P., Castellucci, M., Erbengl, T. and Katowski, A., 1989, Fetal vasculogenesis angiogenesis in human placental villi. *Acta Anatomica* 136:190-203.

Hertig, E., 1935, Angiogenesis in the early human chorion and in the primary placenta of the macaque monkey. *Contri. Embryol. Carnegie Inst.* 25:37-81.

DETECTION OF ANTI-ENDOTHELIAL ANTIBODIES BY WESTERN BLOTTING - POSITIVE CORRELATION WITH CORONARY ARTERY DISEASE: AFTER CARDIAC TRANSPLANTATION

M.J. Dunn, S.J. Crisp, Marlene L. Rose
Patricia M. Taylor and Professor Magdi H. Yacoub

Department of Transplant Immunology
National Heart and Lung Institute
Harefield Hospital
Harefield, Middlesex
United Kingdom

Accelerated coronary artery disease is the most serious complication following cardiac transplantation. The disease has a multi-factorial aetiology, with little agreement about the relative importance of the various risk factors. Here we have investigated the occurrence of anti-endothelial antibodies against human umbilical vein endothelial cells using one-dimensional sodium dodecyl sulphate polyacrylamide gel electrophoresis (SDS-PAGE) and Western blotting. Peptide specific anti-endothelial anti-bodies were found in 15/21 heart transplant recipients with accelerated coronary artery disease, and 1/20 transplant patients who have not developed the disease. Positive immunofluorescence of patients' serum on frozen sections of coronary vessels confirmed the endothelial specificity of antibodies. These results provide evidence of an immune aetiology for transplant-associated coronary artery disease which could have important implications for its diagnosis and therapy.

THE ROLE OF THE $\alpha^V B_3$ INTEGRIN IN SMOOTH

MUSCLE CELL MIGRATION

Leslie Engel, Eric Choi, Kay Broschat, Chris Gorka
Una Ryan and Allan Callow

Health Sciences Department
Monsanto Company
St. Louis, MO 633167
U.S.A.

Vascular Biology Research Laboratory
Department of Surgery
School of Medicine
Washington University
St. Louis, MO 63167
U.S.A.

Intimal hyperplasia is a key event in advanced atherosclerotic lesions. Thickening of the intimal layer of large muscular arteries may result from smooth muscle cells (SMC) that migrate from the muscular wall to the subluminal area and proliferate. The molecular basis for SMC motility is not known.

An assay using transwell tissue culture inserts and a human arterial smooth muscle cell (SMC) line was used to study migration *in vitro*. Several cytokines were tested individually for their ability to stimulate SMC motility; the cytokines included platelet derived growth factor (PDGF), transforming growth factor-β, (TGF-β) vascular permeability factor (VPF), and endothelin - 1 (ET-1). PDGF and ET-1 stimulated migration of SMC in a dose-dependent manner (10ng/ml, 25 ng/ml, respectively) during a 20 hour incubation. No increase in final cell number was observed for any of the cytokines. The concomitant addition of PDGF and ET-1 did not enhance migration. TGF-β or VPF did not stimulate migration beyond background levels.

The integrin complex $\alpha^V \beta_3$ has a role in tumor cell migration (*Seftor et al.,* *1992*). Indirect immunofluorescence and rabbit polyclonal antibodies against $\alpha^V \beta_3 / \beta_5$ (Telios Pharmaceuticals) revealed that SMC express $\alpha^V \beta_3$ on the lower external surface in the form of adhesion plaques. Redistribution of $\alpha^V \beta_3 / \beta_5$ in focal adhesions was observed on PDGF-stimulated SMC. Cells grown in the presence of TGF-β, a combination of PDGF/TGF-β, or in the absence of cytokine revealed that focal adhesions were evenly distributed over the bottom surface of SMC. The results suggest that relocalization of $\alpha^V \beta_3 / \beta_5$ on the lower cell surface of SMC may be necessary for PDGF-or ET-1-induced migration.

The role of the $\alpha^V \beta_3$ integrin complex in cytokine-induced smooth muscle cell migration was examined. The ability of two antagonists of $\alpha^V \beta_3$ to block SMC migration was tested: (1) LM609 (Chemicon, Inc.), a monoclonal antibody that functionally blocks the $\alpha^V \beta_3$ integrin complex; and (2) a RGD peptide antagonist specific for $\alpha \beta^V_3$ (G*penGRGDSPC*A; Telios Pharmaceuticals). LM609 ascites decreased SMC migration in a dose dependent manner. No effect of proliferation or release of cells from the substratum was detected. An isotype matched nonimmune ascites had no effect on motility. In a second experiment, GpenGRGDSPCA was tested for its ability bo block SMC migration. A non functional peptide (GRGESP; Telios Pharmaceuticals) was used as a control. GpenGRGDSPCA inhibited migration in a dose dependent manner. At 10ug/ml GpenGRGDSPCA effectively decreased migration to background values. The percentage motility for GRGESP was 50% for all doses tested. There were no striking differences in final cell number at any dose for either peptide. Indirect immunofluorescence and antibodies against $\alpha^V\beta_3/\beta_5$ suggest that peripheral localization of $\alpha^V\beta/\beta_5$ at sites of focal adhesion may be necessary for PDGF-induced migration. The fact that LM609 antibodies and the GpenGRGDSPCA peptide decreased SMC migration suggests a role for $\alpha^V\beta_3$ in PDGF-stimulated motility.

REFERENCES

Seftor, R.E.B., Seftor, E.A., Gehlsen, K.R., Stetlerstevenson, W.G., Brown, P.D., Ruoslahti, E., and Hendrix, M.J.C. , 1992, Role of the alphavbeta3 integrin in human melanoma cell invasion. *Proc. Natl. Acad. Sci. U.S.A.* 89:(5) 1557-1561.

IRON-SULPHUR NITROSYL COMPLEXES: HIGH CAPACITY,

PHOTOSENSITIVE NO-GENERATORS THAT TARGET VASCULAR

ENDOTHELIUM

F.W. Flitney, I.L. Megson, J.L.M. Thomson, G. Kennovin and A.R. Butler

Cancer Research Group
School of Biological and Medical Sciences
University of St. Andrews
St. Andrews, Fife
Scotland KY16 9TS

Iron-sulphur cluster nitrosyls (A: tetranitrosyltetra-u3-sulphidotetrahedro-tetrainron and B: heptanitrosyl-u3-thioxotetraferrate (1-) spontaneously decompose in solution to generate nitric oxide (NO·). Segments of rat isolated tail artery (7-10 mm) were perfused internally (2ml·min⁻¹) with Krebs' solution containing phenylephrine (4-10µM). Experiments were performed in the dark. Two protocols were used:

First, perfusion pressure was recorded in response to clusters microinjected (10µl) into the internal perfusate. Injection doses less than a critical threshold concentration (D_T) evoked fully-reversible, *transient* (or *T*-type) responses, resembling those seen with conventional nitrovasodilators. Doses $>D_T$ produced long-lasting, *sustained* (or S-type) responses, comprising and initial, rapid drop of pressure, which then either failed to recover or showed partial recovery only. S-type responses were characterized by a stable plateau of reduced pressure which persisted for several hours. Both S- and T- type responses were inhibited by ferro-hemoglobin (Hb; 15µM) and by methylene blue (MB; >10µM) but not by inhibitors of nitric oxide synthase. Addition of Hb or MB to the internal perfusate during an established *S*-type response promptly restored all agonist-induced tone; subsequent removal of Hb caused arteries to re-dilate to the plateau level. These observations demonstrate that the *S*-type response is maintained by slow release of NO· from a molecular "store" in the vessel wall. Histochemical staining confirmed the presence of iron (derived from the clusters) in endothelial cells of treated (but not control) arteries. We conclude that the *S*-type response is the sum of two vasodilator components: a reversible component, due to NO· present in solution at the time of injection: superimposed on a "non-recoverable" component, produced by the spontaneous decomposition of clusters within the endothelium.

Second, pressure recordings were made from arteries perfused continuously with solutions of compound *B* (0.3 or 0.9 µM), initially in the dark, and then during intermittent laser irradiation of the solution *en route* to the preparation λ = 457.9 or 514.5 nm). This generated an additional vasodilatation (light-induced vasodilator response, or LIVR) during the period of illumination. The amplitude of the LIVR was related to the intensity and wavelength of the incident light. The efficiency of photolysis was measured under conditions which precisely simulated those used for recording LIVRs, by collecting exposed solutions and monitoring the change in absorbance at 360 nm. The efficiencies of NO· formation were used to construct log dose-response curves relating LIVR amplitude to [NO·]. A correction was made to allow for loss of NO· in transit from the site of photolysis to the vessel (time delay = 45 sec; $T_{1/2}$ for NO· measured in our apparatus = 33 sec). The results show that half-maximal LIVRs are produced by 15.6 nM NO·.

Our results show that iron-sulphur cluster nitrosyls (a) are potent vasodilators; (b) that they target endothelial cells; and (c) that they are photodegradable. *(This work is supported by the Cancer Research Campaign.*

MIDDLE T TRANSFORMED ENDOTHELIAL CELLS AS A MODEL FOR

VASCULAR TUMORS

Cecilia Garlanda, A. Vecchi, M. Sirone, R. Wainstok de Calmanovici
M. De Rossi, F. Colotta, *C. Parravicini and A. Mantovanni

Institute Ricerche Farmacologiche "Mario Negri"
Milano
Italy

*Ospedale Luigi Sacco
Milano
Italy

In an effort to develop murine models for the study of vascular cell biology and vascular tumors we have investigated a series of Polyoma middle T (PmT) transformed endothelioma cell lines obtained in this and other groups.

Endothelioma cell lines established from different organs of outbred mice inoculated with a retroviral vector containing the PmT oncogene have been reported in the literature (Cell 52: 121, 1988). These lines express PmT antigen, have cobblestone endothelial-like morphology, express Von Willebrand factor, respond to and produce cytokines and chemotactic factors (*J. Immunol.* 147:2122, 1991), cause hemangiomas *in vivo* and have higher fibrinolytic activity compared with normal endothelial cells (EC) (*Cell* 62:435, 1990). The *in vivo* vascular lesions are formed by recruitment of host cells (*Cell* 57:1053, 1989).

Aims of our study were to generate novel PmT transformed EC lines in inbred mice; characterize their biological properties and cytokines production; define the role of interaction with the host (immunity and recruitment) in the pathogenesis of vascular tumors.

We generated EC lines infecting embryonal cells of different tissues of C57Bl/6CrlBR (B6) with the retroviral vector N-TKmT, carrying the PmT oncogene. Stable lines were obtained from whole embryo (line E10V), heart (H5V), brain (B9V).

PmT transformed EC lines are tumorigenic and metastatic in nude and in syngeneic normal mice and cause cavernous hemangiomas. In nude mice $\geq 4 \times 10^4$ H5V cells kill 100% of the hosts. In syngeneic normal B6 mice $\geq 5 \times 10^6$ H5V cells kill 100% of the hosts. Lower doses (10^6, 2×10^5) are rejected respectively by 20% and 70% of normal B6. If B6 mice are irradiated with 450 R 24 h before tumor injection, the percentage of mice rejecting the tumor decreases. Normal B6 that rejected the tumor are resistant to a subsequent inoculum of a lethal dose of H5V, but not of the unrelated syngeneic tumor B16 melanoma.

Metastis are mainly found in the spleen, liver, genital organs and, in the nude mice, skin. The main cause of death is internal hemorrhages for the rupture of the visceral tumors.

Histologically the tumor is formed by vascular cavities filled with blood and fibrin, and separated by connectival stroma highly infiltrated by host cells. The tumoral EC cover the inner wall of the cavities.

Preliminary results obtained by Southern analysis show that only a small part of the tumor mass is composed of PmT transformed cells, suggesting that recruitment of host cells is important in the formation of the vascular lesions.

The PmT transformed EC lines generated in these studies may represent a unique model for the biology and pharmacology of vascular tumors in immunodeficient host.

REGULATORY EFFECTS OF HEPARIN ON LYMPHOCYTE:

ENDOTHELIAL: EXTRACELLULAR MATRIX INTERACTIONS

A. Gorski, B. Dybowska, I. Wojciechowska, M. Nowaczyk

Department of Immunology
Transplantation Institute
Warsaw Medical School
02006 Warsaw
Poland

Recent data emphasize the immunomodulatory effects of low dose heparin (1). It is suggested that the agent alters the traffic of lymphocytes, preventing their migration to a site of antigen, although the precise mechanism of action remains obscure (2). In the first clinical trial in humans we have demonstrated that low doses of heparin, devoid of anti-coagulant effects, inhibit the progression of chronic renal allograft rejection (3). To shed more light on the mechanism of action of heparin at the level of the vessel wall, we studied its effects on lymphocyte interactions with endothelium and the extra cellular matrix (EMC) proteins. Human umbilical vein endothelial cells (HUVEC) were isolated and cultured using standard techniques and used fresh at passage 3 or less. Resting human T cells were isolated using nylon wool columns. Adhesion of ^{51}Cr-labelled T cells and HUVEC to microliter wells coated with ECM proteins (collagen I and IV, fibronectin - CI, CIV, FN) was assessed by incubating the cells for 30 min and subsequent washing of non-adherent cells. Cell binding to protein-coated wells or confluent HUVEC monolayers was assessed by gamma counting or by visual inspection. Low concentrations (below 1 μg/ml) of standard and low molecular weight heparin caused a significant increase in initial attachment of HUVEC to CI and CIV, which probably reflects heparin-mediated enhancement of the binding of HUVEC-associated FN to C (an effect blocked by cycloheximide and anti FN antibodies). In contrast, those doses of heparin inhibit resting T cells adhesion to HUVEC and activated T cells adhesion to CIV. Our data suggest that heparin maintains the integrity of the vessel wall and protects it from T cell-mediated injury

REFERENCES

Gorski, A., Wasik M., Nowaczyk, M. et al., 1991, Immunomodulating activity of heparin. *FASEB J.* 5:-2287.

Lider O., Baharav Y., Mekori, Y. et al., 1989, Suppression of experimental autoimmune diseases and prolongation of allograft survival by low dose heparin. *J. Clin. Invest.* 83:752.

Gorski, A., Leo, M., Gradowska L. et al., 1991, New strategies of heparin treatment used to prolong allograft survival. *Transplant Proc.* 23:2251.

ENDOGENOUS PRODUCTION OF SUPEROXIDE BY THE LUNG: A POSSIBLE ROLE IN ISCHEMIA-REPERFUSION INJURY

Gail H. Gurtner

New York Medical College
Valhalla, NY 10595
U.S.A.

We find spontaneous light emission from isolated Krebs Henseleit perfused rabbit lungs when the light emitting superoxide trap lucigenin is added to the perfusate. Lucigenin light emission appears to be specific for superoxide anion since light emission from the lung caused a superoxide generating system is completely abolished by superoxide dismutase but not catalase of DMTU. We also studied the relative sensitivity of lucigenin photoemission to superoxide and to hydrogen peroxide in vitro. Lucigenin photoemission is three to four orders of magnitude more sensitive to superoxide than to hydrogen peroxide and probably cannot detect hydrogen peroxide in concentrations thought to occur in biologic systems. Basal lucigenin photoemission by the lung is O_2 dependent, since severe hypoxia completely inhibits light emission. Superoxide dismutase reduces photoemission by 50% and administration of the small molecular weight superoxide scavenger tiron inhibits basal photoemission by about 90%. These observations suggest that endogenous superoxide production is primarily intracellular and that approximately half of the dynamically produced superoxide reaches the extracellular space. Superoxide transport may involve anion channels since the anion channel blocker DIDS markedly increases photoemission, suggesting intracellular accumulation of superoxide. We also investigated whether the mechanism for the basal superoxide anion production involved cytochrome P-450 or mitochondrial respiration by using inhibitors of these reactions. The cytochrome P-450 inhibitor, SKF 525A or the mitochondrial transport inhibitor antimycin decreased photoemission by about 50%. These observations suggest that cytochrome P-450 mediated reactions and perhaps mitochondrial function contribute to basal superoxide production in the isolated perfused lung. Endogenous superoxide production may be important in regulation of pulmonary vascular reactivity and may contribute to the pathogenesis of lung reperfusion injury.

SUPEROXIDE MEDIATES HUMAN ENDOTHELIAL CELL

DAMAGE BY STIMULATED GRANULOCYTES

M.M. Hardy, A.G. Flickinger, *T.P. Misko, **R.H. Weiss and U.S. Ryan

Health Sciences Division
Monsanto Company
St. Louis, MO 63167
U.S.A.

*Molecular Pharmacology Division
Monsanto Company
St. Louis, MO 63167
U.S.A.

**Chemical Sciences Division
Monsanto Company
St. Louis, MO 63167
U.S.A

The events that lead to irreversible neutrophil-mediated cell damage are poorly understood. Evidence indicates that the mechanism of injury varies with regard to cell type and the means of neutrophil (PMN) stimulation. The molecular basis of neutrophil-mediated cell injury and inhibitors of this process were evaluated in two *in vitro* assays: a neutrophil-mediated cytotoxicity assay, which measures the release of endothelial bound radiolabel and detects both cell injury and detachment; and a cytochrome *c* reduction assay which measures neutrophil-derived superoxide (O_2-) activity. Human aortic endothelial (HAE) and human dermal microvascular endothelial (HDME) cells were found to differ in susceptibility to neutrophil-mediated injury. The amount of HAE cell injury correlated well with the amount of O_2- generated by stimulated PMN. HDME cells were injured in the presence of lower concentrations O_2-than HAE cells and by PMN stimulated with inflammatory mediators (LPS, PAF, and fMLP) which appeared to generate little or no O_2-. Neutrophil-mediated injury to both HAE and HDME cells was found, however, to be inhibitable by superoxide dismutase (SOD) in a dose dependent manner. The injury was not abrogated by catalase of the iron chelator, deferoxamine mesylate (desferol). When the amount of nitric oxide (NO) released from endothelial cells and unstimulated PMN (measured as nitrite) was examined, HDME cells and PMN were found to release significantly more NO than HAE cells. Stimulation of PMN to generate O_2- resulted in significant decreases in basal nitrite levels. Furthermore, induction of NO synthesis by bradykinin or addition of exogenous NO significantly enhanced neutrophil-mediated HAE cell injury. This injury was abolished by SOD. These results indicate that the superoxide anion is directly involved in the neutrophil-mediated killing of HAE and HDME cells. These data also suggest a role for peroxynitrite (formed upon interaction of NO and O_2-) in oxidative damage. Differential release of NO from cell types may explain variation in susceptibility to neutrophil-mediated injury.

MATRIX METALLOPROTEINASES (MMP's): ROLE IN ATHEROGENESIS AND PLAQUE FISSURE

*W. Hornebeck, **Y. Legrand, ***J. Chapman, *M.P. Jacob
**S. Menaschi and *G. Flores Delgado

*URA CNRS 1460
University of Paris XII
Créteil
France

**INSERM U 353
Hôspital St. Louis
Paris
France

***INSERM U 321
Hôspital La Pitié-Salpétrière
Paris
France

Matrix metalloproteinases (MMP's) a group of neutral zinc endopeptidases, have the capacity to completely degrade all extracellular matrix macromolecules and are directly involved in tissue remodelling. Interactions between human platelets and porcine aorta endothelium cells (EC), which induced an EC shape change, also led to the activation of neutral proteinases: a serine proteinase (M.W. 85 kDa) active on casein gels, designated as PECAP (for "platelet endothelial cells activated protease") and both 70 and 92 kDa Gelatinases (MMP2 and MMP9) two MMP's which possess elastinolytic activities.

The alteration and/or interruptions of the internal elastic lamellae (IEL) has been considered as a main determinant in atherogenesis. Fragmentation of IEL occurs to different extents in different rat strains and was accompanied with increased levels of latent elastase activity secreted by EC. Elastin peptides generated by elastic fibers degradation by elastases possess important biological activities. They may influence cell proliferation and differentiation, modify ion fluxes and stimulate release of elastinolytic enzymes from leucocytes and mesenchymal cells. All these effects are mediated via an elastin receptor sharing similarities with the 67 kDa laminin receptor. They are also potent chemoattractants for monocytes. Incubation of monocytes macrophages with acLDL led to the stimulation of the secretion of elastase activity; this variation in enzyme activity paralleled the increase of intracellular level of cholesterol. Such an effect could be reversed by incubation of cells with HDL3. It was recently evidenced by in situ hybridization that expression of stromelysin in atherosclerotic plaques was mainly localized within large clusters of macrophages that contained intracellular deposits.

It is therefore proposed that MMP's could play a major role in early and late events of atherosclerosis. Thus, inappropriate control by their natural inhibitors, e.g. TIMPs could be responsible for plaque fissure.

REFERENCES

Groult, V., Hornebeck, W., Ferrari, P., Tixier, J.M., Robert L., Jacob, M.P., 1991, Mechanisms of interractions between human skin fibroblasts and elastin:differences between elastic fibers and derived peptides. *Cell Biochem. Funct.* 9:171-182.

Henney, A.M., Wakeley, P.R., Davies, M.J., Foster, K., Hembry, R., Murphy, G. and Hemphries, S., 1991, Localization of stromelysin gene expression in atherosclerotic plaques by *in situ* hybridization. *Proc. Natl. Acad. Sci. USA* 88:8154-8158.

Menashi, S., Hornebeck, W., Robert L., Caen, J., Legrand, Y., 1989, Interactions of platelets with endothelial cells: activation of a novel neutral protease. *Exp. Cell Res.* 183:294-302.

Osborne-Pellegrin, M.J., Farjanel, J., Hornebeck, W., 1990, Role of elastase and lysyl oxidase activities is spontaneous rupture of internal elastin lamina in rats. *Arteriosclerosis* 10:1136-1146.

Rouis, M., Nigon, F., Lafuma, C., Hornebeck, W., Chapman, J., 1990, Expression of elastase activity by human monocytes macrophage is modulated by cellular cholesterol content, inflammatory mediators and phorbol myristate acetate. *Arteriosclerosis* 10:246-255.

THE INHIBITORY ACTION OF LDL AND OXIDIZED ON RELAXATIONS EVOKED BY ENDOTHELIUM-DERIVED NITRIC OXIDE: THE INFLUENCE OF ANTIOXIDANTS

Michael Jacobs, F. Plane and K. R. Bruckdorfer

Department of Pharmacology
Royal Free Hospital of Medicine
London, England
United Kingdom

Department of Biochemistry
Royal Free Hospital of Medicine
London, England
United Kingdom

Hypercholesterolemia and atherosclerosis has been found to impair the vasodilation mediated by the release of endothelium-derived nitric oxide in human and animal arteries. We have previously found that native low-density lipoproteins (LDL) from healthy human donors inhibit acetylcholine (ACh) evoked relaxations in intact rabbit aortic rings precontracted with noradrenaline of serotonin *in vitro*. Cu^{2+}-oxidized LDL inhibits relaxations after preincubation for 30 min and the potency and reversibility of the inhibition depends on the donor (*Jacobs et al., 1990; Plane, F., et al., 1992*). Thus, native LDL from donors whose lipoproteins oxidize rapidly (fast oxidizers) inhibit potently and sometimes irreversibly whereas those which oxidize slowly (slow oxidizers) inhibit less potently and reversibly (*Plane F., et al., 1992*). We now show that antioxidants influence these effects. native LDL (2 mg protein/ml) caused a reduction in sensitively to ACh which was prevented by the addition of 10 uM probucol or 100 uM ascorbic acid, whereas, superoxide dismutase (20 U/ml) had no effect. This indicates that native LDL inhibit via a process involving free radicals but not superoxide. As expected, the antioxidants had no effect *in vitro* in the inhibition by oxidized LDL. Oxidized LDL prepared from the lipoproteins of selected volunteers (fast oxidizers), exhibited reduced and reversible inhibition of relaxation compared with the potent and irreversible inhibition prior to administration of the drug, resembling that of the slow oxidizers. We include that antioxidants such as probucol attenuate the inhibition of endothelium-dependent relaxation by LDL and may prevent the impairment resulting from hypercholesterolemia and atherosclerosis.

REFERENCES

Jacobs, M., Plane, F. and Bruckdorfer, K.R., 1990, *Br. Pharmacol.* 100: 21-26.
Plane, F., Bruckdorfer, K.R., Kerr, P., 1992, *Br. J. Pharmacol.* 105: 216-222.

178

CELL ATTACHMENT AND SPREADING PROPERTIES OF LAMININ

SUBUNITS FROM NORMAL AND NEOPLASTIC TISSUES

Nicholas A. Kefalides, S.J. Wu and W. Jenq

Connective Tissue Research Institute
University of Pennsylvania
Philadelphia, PA 19104
U.S.A.

Department of Medicine
University City Science Center
Philadelphia, PA 19104
U.S.A.

Earlier studies from this laboratory (*Ohno, M. et al., 1983*) demonstrated the presence of a new laminin subunit which we termed "M" in extracts of normal placental membranes. This subunit could not be detected by immunoblot in laminin isolated from the mouse EHS tumor or in cultures of cells isolated from human neoplasms (*Ohno, M. et al., 1986*). In time, the presence of laminin isoforms was demonstrated in normal tissues having the molecular compositions A, B1, B2 (classical), M, B1, B2 (merosin) (*Ehrig, K. et al., 1990*) and A, S, B2 (synaptic laminin) (*Sanes, J.R. et al., 1990*). Additional studies by Rao et al. (*Rao, C.N. et al., 1990*) demonstrated an altered ratio of B1 to B2 subunits as well as of the mRNAs for the same subunits during increased tumorigenesis of A_9 cells to A_9HT cells. In the present study we compared the cell (HT1080) attachment promoting activity of laminin molecules isolated from normal human placenta, a mouse solid tumor (EHS), a normal mouse cell line (B82) and a tumorigenic cell line (B82HT) derived from the former by serial passage in mice. Laminin at various concentrations (0.1, 0.2, 0.5 and 1.0 μg/well) was used as a substrate. HT1080 cells adhered more efficiently when laminin from B82 cells was used as a substrate (40%, 65%, 82% adhesion), compared to laminin from B82HT cells (23%, 35%, 64% and 72% adhesion, at the same substrate concentrations) and to laminin from EHS tumor (10%, 17%, 18%, 31% and 36% adhesion, at the same substrate concentrations). Spreading of HT1080 cells was maximal with the laminin from B82 cells, intermediate with that from B82HT cells and minimal with that from EHS tumor. In an attempt to understand the structure-function relationship between these two cell lines we measured the steady-state levels of mRNAs for A, B1 and B2 laminin subunits in cultures of B82 and B82HT cells. Slot blot hybridization and quantitative densitometry showed the following ratios of mRNA between B82 and B8HT: for A chain 1:0.9, for M 1:0.6, for B1 1:04 and for B2 1:0.34.

In parallel experiments we compared the cell adhesion promoting activity of laminin isolated from human placenta with that of commercial merosin (isolated from placenta) and from mouse EHS tumor. At concentrations of 0.5, 1.0, 1.6 and 5.0 μg/well laminin from the EHS tumor promoted adhesion of HT1080 cells by 15%, 21%, 23% and 41%, respectively. At the same concentrations, placental laminin gave values of 47%, 69%, 77% and 84% adhesion, respectively, and merosin gave comparable values of 54%, 73%, 79% and 89% adhesion, respectively. Spreading of HT1080 cells was comparable at the low concentrations of laminin from placenta and merosin but very minimal with laminin from EHS tumor. Although differences were noted between the laminin from the EHS tumor and the placenta with respect to adhesion of HT1080 cells, no such differences were observed between the two types of laminin when human umbilical vein endothelial cells were tested. These studies demonstrate that laminin from normal tissue is more efficient in promoting cell adhesion compared to laminin from a tumor and the difference may be cell specific. The single most important difference between these two groups of laminin is the absence of the M subunit in the EHS molecule. The data also suggest that the expression or non-expression of the M subunit of laminin may be related to increased tumorigenicity.

REFERENCES

Ehrig, K., Leivo, I., Argraves, W.S., Ruoslahti, E. and Engvall, E., 1990, Merosin, a tissue-specific basement membrane protein is a laminin-like protein. *Proc. Natl. Acad. Sci. USA* 87:3264-3268.

Ohno, M., Martinez-Hernandez, A., Ohno, N. and Kefalides, N.A., 1983, Isolation of laminin from human placental basement membranes: amnion, chorion and chorionic microvessels. *Biochem. Biophys. Res. Comm.* 112:1091-1098.

Ohno, M., Martinez-Hernandez, A., Ohno, N. and Kefalides, N.A., 1986, Laminin M is found in placental membranes but not in basement membranes of neoplastic origin. *Connect. Tissue Res.* 15:199-207.

Rao, C.N., Brinker, J.M. and Kefalides, N.A., 1990, Changes in the subunit composition of laminin during the increased tumorigenesis of mouse A_g cells. *Connect. Tissue Res.* 25:321-329.

Sanes, J.R., Engvall, E., Butkowski, R. and Hunter, D.D., 1990, Molecular heterogeneity of basal laminae: Isoforms of laminin and collagen IV at the neuromuscular junction and elsewhere. *J. Cell Biol.* 111:1685-1699.

ENDOTHELIN IN THE ADULT RESPIRATORY DISTRESS SYNDROME

David Langleben, Michel DeMarchie and Duncan Stewart

Sir Mortimer B. Davis Jewish General Hospital and
Royal Victoria Hospital
McGill University
Montreal, Quebec H3T 1E2
Canada

The Adult Respiratory Distress Syndrome, ARDS, is a major cause of morbidity and mortality in the intensive care setting. ARDS can be caused by a variety of illnesses, including systemic sepsis, trauma and burns. However, all form of ARDS have in common an initial pulmonary endothelial injury, leading to pulmonary edema and respiratory failure. Later activation of secondary mediators causes thrombosis and pulmonary vascular muscularization, which narrows the pulmonary bed, resulting in pulmonary hypertension. Many of the mediators of this process are as yet unknown. Endothelin-1 (ET-1) is recently described vasoconstrictor peptide which has potent smooth muscle-mitogenic activity *in vitro*. Through these actions, ET-1 might contribute to the progression of vascular narrowing in ARDS. We therefore measured ED-1 levels in patients with ARDS.

The human lung normally clears circulatory ET-1, so that systemic arterial ET-1 levels are normally lower than mixed venous ET-1 levels entering the lung, and the mean arterial to venous ratio is normally 0.59 ± 0.12. We measured plasma ET-1 levels in mixed venous blood (Venous), from a Swan Ganz catheter, and systemic arterial blood (arterial) from an arterial line. ET-1 was measured by RIA. Thirty-one patients with ARDS were examined. Blood was sampled at a mean of 1.6 ± 0.31 SEM days after onset of the respiratory failure. The ARDS patients had elevated plasma ET-1 levels (4.32 ± 0.49 pg/ml venous, and 4.66 ± 0.5 pg/ml arterial) as compared to normal controls (1.45 ± 0.16 pg/ml and 0.92 ± 0.25 pg/ml, respectively). In addition, the arterial/venous ET-1 ratio was 1.2 ± 0.1, implying net release of ET-1 from the lung. Sixteen of the 31 patients died, and there were no differences in pulmonary artery pressure, total pulmonary vascular resistance, venous ET-1, arterial ET-1 or arterial/venous ratio between the groups that lived or died.

We obtained a second set of blood samples in 10 of the above patients, at a time of deterioration of ventilatory status (mean 6.7 ± 1.5 days after 1st sample) and in 12 of the above patients at a time of improvement of ventilatory status (mean 31.5 ± 18.1 days after 1st sample).

In the group that improved, mean pulmonary artery pressure fell from 29 ± 1.4 mmHg to 24.8 ± 2.2 mmHg (p < 0.02) and total pulmonary vascular resistance from 5.9 ± 0.7 Wood units to 4.35 ± 0.31 Wood units. Similarly, venous ET-1 levels fell from 3.8 ± 0.6 pg/ml to 1.7 ± 0.2 pg/ml (p < 0.01) and arterial ET-1 levels fell from 3.8 ± 0.5 pg/ml to 1.5 ± o.3 pg/ml. In the recovery group, the arterial/venous ratio fell from 1.1 ± 0.1 to 0.8 ± 0.1 (p < 0.05). The group that worsened continued to exhibit severe abnormalities in hemodynamics, and persistent elevations in plasma ET-1 levels and arterial/venous ratio.

Thus, in patients with ARDS, circulating venous and arterial plasma ET-1 levels are markedly elevated by day 1 of illness, and the arterial/venous plasma ET-1 ratio suggests release of ET-1 by the lung at this earl;y stage of ARDS. Furthermore, these abnormalities remain abnormal in patients who have ongoing ARDS, but they improve in patients with ARDS who recover from ARDS.

Endothelin-1 release and activity in patients with ARDS may contribute to the vascular abnormalities seen in this disorder. (*Supported by Canadian Heart Foundation, Fonds de la Recherche en Sante du Quebec, and MRC Canada.*)

THAPSIGARGIN DIFFERENTIATES BETWEEN EDRF AND PROSTACYCLIN

RELEASE INDUCED BY SHEAR-STRESS OR AGONISTS

Heather Macarthur, M. Hecker and J.R. Vane

The William Harvey Research Institute
St. Bartholomew's Hospital Medical College
Charterhouse Square
London EC1M 6BQ
United Kingdom

The agonist-mediated simultaneous release of endothelium-derived relaxing factor (EDRF) and prostacyclin (PGI_2) is initiated by a mobilization of intracellular calcium ($[Ca^{2+}]_i$). It is, however, as yet unclear how this increase in $[Ca^{2+}]_i$ relates to the physiologically more important, release of EDRF induced by shear stress. Recently the tumor-promoting sesquiterpene lactone, thapsigargin, was shown irreversibly to inhibit the re-uptake of CA^{2+} into inositol-1,4,5-triphosphate ($InsP_3$)-sensitive Ca^{2+} stores by blocking the Ca^{2+}-ATPase located in the endoplasmic reticulum of various cells, including platelets and polymorphocyclear cells (*Thastrup et al., 1990, Takemura et al., 1991*). By employing this Ca^{2+}-ATPase inhibitor, we have now investigated further the effects of inhibition of intracellular Ca^{2+} sequestration on the release of EDRF and PGI_2 from bovine aortic endothelial cells (BAEC). The cells (grown on microcarrier beads with 2 ml corresponding to 6×10^7 BAEC) were packed into a jacketed chromatography column and perfused (5 ml/min) with warmed (37°C), oxygenated (95% O_2/5% CO_2) Kreb's solution. The effluent superfused an endothelium denuded rabbit aortic ring preconstricted to 2-3g tension with 10 nM U46619 (9α, 11α-methaneoepoxy-prostaglandin F_{2a}) for the detection and quantification of EDRF release. PGI_2 release was determined by radioimmunoassay for 6-keto-prostaglandin F_{1a} and changes in $[Ca^{2+}]_i$ were monitored with Fura-2/AM loaded BAEC in suspension. EDRF release induced by shear stress was defined as the degree of relaxation (16.7 ± 3% of induced tone; n=5) caused by an infusion of superoxide dismutase (SOD; 10 U/ml) through the column of BAEC (t.c). This relaxation was further enhanced (10-25%; n=5) by bolus injections of agonists such as ADP (9 nmol), ionomycin (60 pmol), or poly-L-lysine (550 pmol). Infusions t.c of thapsigargin (1 μM) caused a sustained release of EDRF (29.8 ± 8.1% relaxation; n=5) which declined after stopping the infusion. Thereafter, the agonist-induced release of EDRF was abolished, whereas, the shear stress-induced release remained unaffected (n=5). The EDRF activity after both types of release was blocked by infusions t.c of hemoglobin (10 μM; n=3). The basal release of PGI_2 from the BAEC (1.04 ± 0.27 ng/min; n=5) was significantly enhanced (3-6 ng/min) by bolus injections t.c of any of the three agonists. Infusions of thapsigargin t.c also caused a substantial increase in PGI_2 release (5.04 ± 1.99 ng/min), but thereafter, as with EDRF, challenges by the agonists failed to cause any further rise in PGI_2 release. The resting $[Ca^{2+}]_i$ level in Fura-2/AM-loaded BAEC (5×10^5 cells/ml) was 70-100 nM. ADP (1 μM) produced a typical biphasic change in $[Ca^{2+}]_i$ i.e a marked transient (< 2 min) rise (peak 400-1000 nM) followed by a plateau phase (200-300 nM) lasting for more than 10 min (n=4). Thapsigargin

(1 μM) elicited an initial sharp increase in $[Ca^{2+}]_i$ (Peak 800-1200 nM) which declined slowly and remained elevated throughout the experiment at approximately 500 nM. Subsequent exposure to ADP produced no further increase in $[Ca^{2+}]_i$ (n=4). It appears, therefore, that by emptying the $InsP_3$-sensitive and insensitive Ca^{2+} pools, thapsigargin selectively blocks the agonist-stimulated release of EDRF and PGI_2, suggesting a fundamental difference in the control by CA^{2+} of release of both substances by shear stress or agonists. (*Supported by a grant from Glaxo Group Research Ltd.*)

REFERENCES

Takemura, H. et al., 1991, *Biochem. Biophys. Res. Commun.* 180:1518-1526.
Thastrup, O. et al., 1990, *Proc. Natl. Acad. Sci. U.S.A.* 87:2466-2470.

L-GLUTAMINE TRANSPORTER IN CULTURED HUMAN UMBILICAL VEIN ENDOTHELIAL CELLS: CHARACTERISTICS AND ROLE IN OXIDANT STRESS

Lisa A. Madge, A.R. Baydoun, J.D. Pearson and G.E. Mann

Vascular Biology Research Center
Biomedical Sciences Division
King's College
Campden Hill Road
London W8 7AH
United Kingdom

The vascular endothelium may be subject to oxidative stress and injury in the pathogenesis of many diseases including inflammation and atherosclerosis (*Morel et al., 1984*). Sublytic injury of the endothelium is associated with a widespread change in cellular function including changes in vascular permeability (*Block, 1991*), depletion of ATP and inhibition of glycolysis (*Hyslop et al., 1988*). Availability of extracellular L-glutamine may act as a substrate for ATP generation and thus protect against oxidative stress (*Hinshaw et al., 1990*). In addition, L-glutamine regulates the transport of L-cystine (*Bannai and Ishii, 1988*) and may have a further protective role in oxidant stress by maintaining levels of intracellular Glutathione. We have characterized the transport of L-glutamine in human umbilical vein endothelial cells (HUVEC) and performed preliminary studies to set up a model of oxidative stress in this cell type.

Endothelial cells were grown to confluency in either 96 well or 6 well plates containing M199 medium supplemented with 4mM L-glutamine, 10% fetal and 10% newborn calf serum. Transport studies were carried out in HEPES (20mM) buffered Krebs' solution (37^0C) and initiated by the addition of [^{3}H]glutamine (tracer plus 100μM cold L-glutamine) to cell monolayers. Oxidative stress was induced by the addition of H_2O_2 (50-2mM) to each well and release of lactate dehydrogenase (LDH) determined 0.5 - 24 hours after the addition of the oxidant. Depletion of intracellular ATP was determined after 15 minutes by High Performance Liquid Chromatographic analyses.

Uptake of L-glutamine was found to be linear for up to 20 minutes and occurred against a marked concentration gradient. Kinetic analysis of L-glutamine transport (0.05 - 1mM) revealed saturable kinetics with a V_{max} of 12 ± 1 pmol/min/μg protein and K_m of 150 ± 22 μM. Uptake was inhibited in the absence of Na^+($46 \pm 6\%$ of control, n = 11) and was not restored to control values with lithium substitution. The transporter was also inhibited by L-asparagine, L-serine, L-cysteine and to a lesser extent 2-methylaminoisobutyric acid. L-histidine, β-2-amino-bicyclo-(2,2.1)heptane-2-carboxylic acid (BCH) and L-arginine were ineffective inhibitors. This data in conjunction with the insensitivity of the transporter to pH suggest that L-glutamine transport is largely mediated by System ASC as described in pulmonary artery endothelium (*Herskowitz et al., 1991*).

Preliminary data inducing oxidative stress shows that incubation of endothelial cells with H_2O_2 for 15 minutes does not result in the release of LDH. Intracellular ATP however, shows a marked depletion with concentrations of $H_2O_2 \geq 250\mu M$ and may, therefore, provide a more sensitive index of oxidative stress. The protective role of L-glutamine in this system remains to be established.

REFERENCES

Bannai, S. and Ischii, T., 1988, *J. Cell Physiol.* 137:360-366.

Block, E., 1991, *J. Cell. Physiol.* 146:363-393.

Herskowitz, K., Bode, B., Block, E. Souba, W., 1991, *Am. J. Physiol.* 248:550-556.

Hinshaw, D., Burger, J., Delius R. and Hyslop, P., 1990, *Surgery* 108:298-305.

Hyslop, P.A., Hinshaw, D.B., Halsey, W.A., Scraufstatter, I.U., Jackson, J.H., Spragg, R.A., Sauerheber, R.D. and Cochrane, C.G., 1988, *J. Biol. Chem.* 263: 1655.

ISCHAEMIA AND REPERFUSION ENHANCES THE VASOCONSTRICTOR EFFECTS OF ENDOTHELIN-1 IN THE ISOLATED PERFUSED HEART OF THE RAT

Lorraine McMurdo, C. Thiemermann and J.R. Vane

The William Harvey Research Institute
St. Bartholomew's Hospital Medical College
Charterhouse Square
London
United Kingdom

Plasma levels of ET-1 (and its precursor big ET-1) are elevated in patients with acute myocardial infarction (*Miyauchi et al., 1989*). Cardiac membranes obtained from isolated perfused hearts of rats subjected to 30 min of global ischemia then reperfusion showed an increased density of ER-1 reduces infarct size after coronary artery ligation and reperfusion in anesthetized rats (*Watanabe et al., 1990*). Here we demonstrate that the vasoconstrictor effects of ET-1 in the coronary vascular bed are enhanced after 30 min of global myocardial ischemia followed by reperfusion. In addition, this study investigates whether a reduced formation of either prostacyclin or nitric oxide accounts for the ischemia-induced enhancement of the vasoconstrictor response to ET-1.

Male Wistar rats (350-450g) were anesthetized with sodium pentobarbitone (75 mg kg^{-1}, i.p.) After injection of heparin (500 IU, i.v.) and thoracotomy, the hearts were excised and perfused (according to Langendorff) at a constant flow of 10 ml min^{-1} with oxygenated (95% O_2, 5% CO_2) Krebs' solution. Changes in coronary perfusion pressure (CPP) were recorded in control (NMI) hearts and in hearts subjected to 30 min of global ischemia (MI) followed by 30 min of reperfusion. Drugs were given as bolus injections 10 min after restoration of flow in MI hearts. Results are expressed as the increase/decrease in CPP (mm Hg) or as the area under the curve in arbitrary units (AU).

ET-1 (100 pmol) caused a rise in CPP (52 ± 4 AU; n=8), which was significantly enhanced after MI and reperfusion (108 ± 10 AU; n=9). Indomethacin (5.6 μM) enhanced the ET-1 induced rise in CPP in NMI hearts (81 ± 8 AU; n=7). Moreover, MI and reperfusion further potentiated the augmentation of the ET-1 pressor response (144 ± 16 AU; n=8). Similarly, L-N^G-nitro arginine methyl ester (NO2-Arg; 30 μM) augmented the ET-1 mediated rise in CPP in NMI hearts (125 ± 21 AU; n=4). However, MI and reperfusion had no effect on the pressor response in the presence of NO2-Arg (120 ± 13; n=6). When compared to ET-1, phenylephrine (1 μmol; n=4) or the thromboxane receptor agonist U46619 (3 nmol; n=6) produced as similar rise in CPP in NMI hearts. In contrast to ET-1, the pressor responses to these agonists were attenuated by MI and reperfusion. NO2-Arg did not significantly alter the constrictor responses to phenylephrine but did significantly increase the responses to U46619 (n=4).

These results demonstrate that global MI followed by reperfusion of the ischemic vascular bed enhances the coronary vasoconstrictor effects of ET-1. This augmentation of the pressor response is unlikely to be due to a reduced formation or release of either prostacyclin or nitric oxide from the vascular endothelium.

REFERENCES

Liu, J. et al., 1990, *J. Cardiovasc. Pharmacol.* 15:436-443.
Miyauchi, T. et al., 1989, *Lancet* 2:53-54.
Watanabe, T. et al., 1990, *Nature* 348:673.

POTENTIATION OF ENDOTHELIUM DIRECTED IMMUNE COMPLEX-MEDIATED CYTOTOXICITY OF NEUTROPHILS BY CYTOKINE ACTIVATION OF THE ENDOTHELIAL MONOLAYERS: A NEW *IN VITRO* MODEL FOR LEUKOCLASTIC VASCULITIS

Rene Moser, L. Olgiati and J. Fehr

Department of Medicine/Hematology
University Hospital
Zurich
Switzerland

Polymorphonuclear neutrophils (PMN), which display intense spreading and metabolic activation on plastic tissue culture dishes (*Dahinden, C. et al., 1983a; Dahinden, C. et al., 1983b*), showed only loose attachment to endothelial cell monolayers (ECML) without any measurable metabolic activation and cytotoxicity (*Fehr, J. et al., 1985*). In the light that endothelial cells stimulated with IL-1 and TNF actively control adherence and transendothelial migration in the absence of a chemotactic gradient (4), metabolic activation of the interacting PMN was of interest. The endothelial-dependent type of adherence is characterized by intense interaction of the PMN membrane with the activated endothelium (*Moser, R. et al., 1989*). Such tight cell contact is, in terms of sheer stress resistance, comparable to the cell substrate interaction of PMN spreading on artificial surfaces which results in severe metabolic activation and degranulation (*Dahinden, C. et al., 1983a; Dahinden, C. et al., 1983b*). However, the endothelial-dependent adherence had no influence on the PMN metabolic activity (*Moser, R. et al., 1989*).

In turn, PMN attaching to ECML via surface bound immune complexes (IC) liberated their cytotoxic potential driven by a marked burst of PMN hexose monophosphate shunt activity (HMPSA) and accompanied by a congruent release of elastase and vitamin B_{12}-binding protein (B_{12}-BP). Successive activation of the monolayer (2h) with IL-1 (5U/ml) or TNF (50 ng/ml), led to an impressive augmentation of HMPSA and enzyme release resulting in enhanced cytotoxicity as measured by the release of ^{51}Cr from labeled ECML. The observed cytotoxic activity was proportional to the number of interacting neutrophils. The cytokine-dependent effect on PMN activation was dependent upon CD11/CD18 and was partially inhibited by mAbs against ELAM-1 and ICAM-1.

From these results, we conclude that the mechanism of cytokine-augmented Fc-mediated cytotoxicity basically relies on the recruitment capacity for PMN of IL-1 and TNF-activated endothelium. This *in vitro* model represents a novel pathophysiologic concept of IC-mediated vasculitis.

REFERENCES

Dahinden, C., Galanos, C. and Fehr, J., 1983, Granulocyte activation by endotoxin. Part I. *J. Immunol* 130:857-862.

Dahinden, C., Galanos, C. and Fehr, J., 1983, Granulocyte activation by endotoxin Part II. *J. Immunol* 130:863-868.

Fehr, J., Moser, R., Leppert, D. and Groscurth, P., 1985, Antiadhesive properties of biological surfaces are protective against stimulated granulocytes. *J. Clin. Invest.* 76:535-542.

Moser, R., Schleiffenbaum, B., P. Groscurth, P. and Fehr, J., 1989, Interleukin 1 and tumor necrosis factor stimulate human vascular endothelial cells to promote transendothelial passage. *J. Clin. Invest.* 83:444-455.

ENDOTHELIAL FUNCTION IN CARDIAC TRANSPLANT RECIPIENTS

Greg S. O'Neil, Adrian H. Chester and Magdi H. Yacoub

Thoracic and Cardiac Surgical Unit
Harefield Hospital
Harefield
Middlesex UB9 6JH
United Kingdom

Objective: It has been hypothesized that toxic manifestations associated with cyclosporin therapy may be contributed to, in part, by compromised endothelial function. *In vitro* animal studies have implicated inhibition of release of the endothelium-derived relaxing factor, nitric oxide; however, this has not been investigated in human tissue. The present study investigates the effect of cyclosporin A on nitric oxide release in human coronary arteries.

Design: This study has employed both *in vitro* organ bath preparations and *in vivo* measurements in the coronary circulation.

Patients: For *in vitro* experiments, coronary arteries were harvested from excised heart of 10 patients requiring transplantation for reasons other than ischemic heart disease. Of these, 3 patients were being re-transplanted for obliterative bronchiolitis and had been receiving cyclosporin for a mean of 22 months.

The *in vivo* study was performed on a group of 12 cardiac transplant recipients who were clinically well 1-5 years post-operatively, and were not undergoing allograft rejection at the time of assessment.

Results: Isolated vessel segments *in vitro* relaxed in a dose-dependent manner to substance P (10^{-11} - 10^{-7}M) with a maximum response of $76.6 \pm 7.4\%$ of the response to 1 ug/ml glyceryl trinitrate. Incubation with 1000 and 2000ng/ml cyclosporin abbreviated this response to $63.0 \pm 11.5\%$ and $62.2 \pm = 11.1\%$ respectively; this was not significant ($p > 0.05$).

In segments taken from explanted hearts of patients requiring re-transplantation, the mean maximum response was $78.0 \pm 11.0\%$, and there was no correlation between maximum response in segments from each patient, and duration of cyclosporin therapy.

The effect of intracoronary substance P in 12 cardiac transplant recipients was also examined (mean cyclosporin blood level: 228.9 ± 42.8 ng/ml). The mean maximum dilation measured as % diameter change induced by substance P and isosorbyldinitrate was $22.1 \pm 3.2\%$ and $26.0 \pm 2.5\%$, respectively. There was no correlation between degree of endothelial-mediated vasodilation to substance P, and cyclosporin level.

Conclusions: We conclude that the nitric oxide response is preserved in coronary arteries of patients exposed to cyclosporin. The mechanisms that initiate cyclosporin-associated toxicity remain to be elucidated.

ASSAY OF PULMONARY MICROVASCULAR ENDOTHELIAL

ANGIOTENSIN CONVERTING ENZYME, *IN VIVO*: COMPARISON

OF THREE METHODS

*S.E. Orfanos, *X-L Chen, **J.W. Ryan, **A.Y.K. Chung,
***S.E. Burch and *J.D. Catravas

*Department of Pharmacology & Toxicology
Medical College of Georgia
Augusta, GA 30912
U.S.A.

**Department of Medicine
University of Miami School of Medicine
Miami, FL. 33101
U.S.A.

and

***Department of Radiology
Medical College of Georgia
Augusta, GA. 30912
U.S.A.

We monitored the activity of pulmonary microvascular endothelial-bound angiotensin converting enzyme (ACE), *in vivo*, by means of multiple indicator-dilution type techniques, utilizing three different probes: the hydrolysis of two substrates, ^{3}H-benzoyl-Phe-Ala-Pro (BPAP) and ^{14}C-benzoyl-Ala-Gly-Pro (BAGP), and the binding of the inhibitor ^{3}H-RAC-X-65 (RAC), all measured during a single transpulmonary passage in anesthetized rabbits, placed on total heart by-pass, so that both systemic and pulmonary circulations were fully supported by means of a two-channel extracorporeal pump. Experiments were performed at pulmonary blood flows ($\dot{Q}_b$) of 250, 400, 560 and 800 ml/min, in control or indomethacin pretreated rabbits. We also investigated the activity of pulmonary microvascular endothelial-bound 5´-nucleotidase (NCT), by measuring the dephosphorylation of its natural substrate ^{14}C-5´-AMP. We calculated substrate utilization, mean lung transit time ($\bar{t}$), and volume of distribution (i.e., central blood volume) of all three substrates, as well as inhibitor binding. We also calculated A_{max}/K_m and B_{max}, the modified parameters for substrates and inhibitor, respectively. As $\dot{Q}_b$ increased, A_{max}/K_m values for all three substrates and B_{max} increased linearly, indicating microvascular recruitment. In experiments where either BPAP and 5´-AMP metabolism, or BAGP metabolism and RAC binding were studied concomitantly, a linear relationship was observed between $\dot{Q}_b$-induced changes in A_{max}/K_m values of BPAP vs 5´-AMP as well as in A_{max}/K_m of BAGP vs B_{max} of RAC. Similarly, increasing $\dot{Q}_b$ increased central blood volume and decreased $\bar{t}$, substrate utilization and inhibitor binding. Indomethacin had no effect on most of the hemodynamic or enzyme parameters measured. We conclude that *in vivo* assays of ACE proceed as predicted by Michaelis-Menten kinetics and offer insights into pulmonary endothelial pathophysiology. (*Supported by HL 31422 and HL 46689*)

RADIATION-INDUCED ALTERATION IN ANGIOTENSIN CONVERTING ENZYME ACTIVITY IN CULTURED BOVINE PULMONARY ARTERY ENDOTHELIAL CELLS

Andreas Papapetropoulos, *S.E. Burch, Stavros Topouzis and John D. Catravas

Department of Pharmacology & Toxicology
Medical College of Georgia
Augusta, GA 30912-2300
U.S.A.

*Department of Radiology
Medical College of Georgia
Augusta, GA 30912
U.S.A.

Bovine pulmonary arterial endothelial cells (BPAE) were exposed to a single dose 0, 5, 10, 20 or 30 Gy. Angiotensin converting enzyme (ACE) activity was determined in confluent monolayers under first order reaction conditions at 6, 24, 48 and 96 hr after treatment using [^{3}H]-benzoyl-Phe-Ala-Pro as substrate. Irradiation decreased the number of viable endothelial cells in a dose- and time- dependent manner beginning at 24 hr after 5 Gy and reaching a maximum effect (20% survival) at 96 hr after 30 Gy. Total amount of protein per monolayer decreased during the same intervals, whereas protein content per cell rose signifying a radiation-induced hypertrophy of viable cells. When expressed per million surviving cells, ACE activity increased in a time- and dose-dependent manner beginning at 24 hr after 5 Gy and reaching a maximum four-fold increase at 96 hr after 30 Gy. "These results suggest that although at the lowest radiation dose (5 Gy), the increase in ACE activity per cell compensated for the enzymatic activity lost due to cell death at higher doses (10, 20 and 30 Gy), the increase in ACE activity per cell could not keep up with decrease in number of viable endothelial cells leading to an overall decrease in total ACE activity per culture well.

"However, when expressed per culture well, ACE activity decreased in a time and modulation-dependent manner.

194

ENDOTHELIUM-DERIVED NITRIC OXIDE SYNTHASE INHIBITION:

EFFECTS ON CEREBRAL BLOOD FLOW, PIAL ARTERIAL DIAMETER AND

VASCULAR MORPHOLOGY IN THE RAT

R. Prado, Brant D. Watson, and W.D. Dietrich

Cerebral Vascular Disease Research Center
Department of Neurology
University of Miami School of Medicine
Miami, FL
U.S.A.

The purpose of this study was to determine the effects of inhibiting endothelial-derived relaxing factor (nitric oxide) on cerebral vascular integrity in the rat by: 1. measuring large pial artery diameter through an operating microscope, 2. monitoring cortical cerebral blood flow (cCBF) by laser-doppler flowmetry and 3. examining morphological changes in the ultrastructure of the vascular endothelial cell. The nitric oxide synthase inhibitor N(ω)-nitro-L-arginine methyl ester hydrochloride (L_NAME), or saline was infused via a retrograde right external carotid artery catheter at a rate of 3 mg/kg/min to a total dose of a 190 mg/kg (n=16), or intravenously at 1 mg/kg/min to a total dose of 15 mg/kg (n=6) or 30 mg/kg (n=8).

Infusion of L-NAME significantly raised arterial blood pressure in all three doses (for 190 mg/kg, 103.2 ± 3.4 to 135 ± 3.4 mmHg; for 15 mg/kg, 125 ± 2.8 to 144.4 ± 4.0 mmHg; for 30 mg/kg, 99.5 ± 2.8 to 138 ± 5.4 mmHg, all observed at 60 min after the start of infusion). The average decrease in pial artery diameter during 1 hr of observation following the intracarotid infusion of L-NAME (3 mg/kg/min) was 39 ± 7.4% (n=4) of preinfusion diameter, while cortical cerebral blood flow decreased to an average of 72.5% of baseline (n=7). Intravenous administration of L-NAME (15 mg/kg, n=6) decreased cCBF to an average of 72.8% of baseline at the end of the 60 min observation period. Thirty minutes following the intravenous infusion of L-NAME (30 mg/kg, n=8) pial artery diameter decreased from 157.3 ± 12.1 to 100.0 ± 10.0 um. Morphological abnormalities observed at 1 hr after infusion with L-NAME (3 mg/kg/min, n=5) included microvascular stasis within cortical and subcortical regions. Ultrastructural examination of cortical vessels revealed constricted arterioles with large numbers of endothelial microvilli projecting into the vessel lumen.

These results suggest that the regulation of blood flow in the cerebral circulation is affected by artificially inhibiting nitric oxide synthase and can lead to subtle alterations in the morphological characteristics of the cerebral vasculature consistent with a pro-thrombotic state.

NERVOUS CONTROL OF ENDOTHELIUM-DEPENDENT

RELAXATION IMPLICATIONS FOR DISEASE STATES

T.M. Scott, Karen Drodge and Judy Foote

Faculty of Medicine
Memorial University of Newfoundland
Newfoundland
Canada

In disease states such as atherosclerosis, hypertension and diabetes there is a reduction in the ability of blood vessels to relax. The EDRF system appears to be involved in some fashion. The change in relaxing ability varies between diseases and between vascular beds. In many animal models of hypertension a depression of EDRF induced relaxation has been shown to occur (*Konishi and Su, 1983; Winquist et al., 1984; De May et al., 1985; Luscher et al., 1987; Miller et al., 1987*). Similar results have been reported in animal models of diabetes (*Durante et al., 1988*), and in atherosclerosis (*Bossaller et al., 1987; Harrison et al., 1987; Jayakody et al., 1988; Guerra et al., 1989*).

In many of these disease states there is in addition to alterations in the EDRF system a concomitant change in perivascular innervation. Studies using capsaicin have suggested a role for perivascular innervation in th control of the EDRF system (*Scott et al., 1992*). We have investigated an alternative model of vascular denervation. A small series of Sprague-Dawley rats was anesthetized and the superior mesenteric arterial bed exposed. Using liquid nitrogen-cooled cotton-tipped sticks, a 2mm section of the superior mesenteric artery was repeatedly frozen and thawed following the method of Hill *et al.*, (1985). At various times after freezing the isolated perfused superior mesenteric arterial bed was examined. It was determined that close to normal relaxation responses to ACh could be elicited at two days after freezing however, there was a reduction in endothelium-dependent relaxation by seven days after freezing, continuing to fourteen days after freezing. By twenty-one days after freezing the responses had returned towards normal. Immunohistochemistry demonstrated a loss of CGRP/SP containing nerve fibers by two days after freezing Electron microscopy demonstrated increasing profiles representing degenerating axons from two to twenty-one days after freezing. Removing perivascular innervation by freezing results in a reduction in endothelium-dependent relaxation confirming the findings following capsaicin induced denervation.

These results have implications for disease states in which perivascular innervation is disturbed and in the case of organ transplant where the tissue is denervated.

REFERENCES

Bossaler et al., 1987, *J. Clin. Invest.* 79:170-174.
De May, J.G. and Gray, S.D., 1985, *Prog. Appl. Microcirc.* 8:181-187.

Guerra, R. et al., 1989, *Blood Vessels* 26:300-314.

Harrison et al., 1987, *Circ. Res. 61 Suppl.* 11:11-80.

Hill, C.E., Hirst, G.D.S., Ngu, M.C. and van Helden, D.F., 1985, *J. Aut. Nerv. Syst.* 14:317-334.

Konishi, M. and Su, C., 1983, *Hypertension* 5:881-886.

Luscher et al., 1987, *Hypertension* 9:157-163.

Miller et al., 1987, *Hypertension* 10:164-170.

Scott et al., 1992, *Artery* 19:221-224.

Winquist et al., 1984, *J. Hypertension* 2:541-545.

VITAMIN E SUPPLEMENTATION, PLASMA LIPIDS AND INCIDENCE OF RESTENOSIS AFTER PERCUTANEOUS TRANSLUMINAL CORONARY ANGIOPLASTY (PTCA)

**Demetrios S. Sgoutas, S. J. DeMaio, S.B. King, III, N.J. Lembo
G.S. Roubin, J.A. Hearn, *H.N. Bhagavan, A. Gruentzig

Cardiovascular Center
Department of Internal Medicine (Cardiology)
Atlanta, GA 30322
U.S.A.

*Hoffman-La Roche Inc.
Nutley, NJ 07110
U.S.A.

**Department of Pathology and Laboratory Medicine
Emory University School of Medicine
Atlanta, GA 30322
U.S.A.

To test whether α-tocopherol prevents restenosis following percutaneous transluminal coronary angioplasty (PTCA) we enrolled patients in a double-blind, placebo-controlled trial. Patients were randomized after successful PTCA to receive vitamin E in the form of dl-α-tocopherol, 1200 IU per day, orally vs. inactive placebo for 4 months. Patients' blood was analyzed at baseline and at 4 month post-PTCA, for differences in plasma lipids, lipoproteins, apolipoproteins, α-tocopherol, retinol, β-carotene and lipoperoxides concentrations. One hundred patients completed the protocol. No significant difference was found in any parameter except the α-tocopherol levels between the vitamin E group and the placebo group, verifying compliance. Follow-up cardiac catheterization was obtained in 83% of the patients receiving placebo and in 86% of the patients receiving dl-α-tocopherol. Including thallium and exercise stress testing, objective information was obtained for practically all the patients receiving dl-α-tocopherol or placebo. Restenosis was defined as the presence of a lesion with $\leq 50\%$ stenosis in a previously dilated artery segment diameter, vessel diameter at follow-up, and restenosis rate. Patients receiving dl-α-tocopherol had a 35.5% restenosis angiographically documented vs. 47.5% restenosis in patients receiving placebo. The overall incidence of restenosis defined by an abnormal angiogram or thallium test or exercise stress test was 34.6% in patients receiving dl-α-tocopherol and 50.0% in patients receiving placebo. This difference (at $p = 0.06$) did not reach significance because of an inadequate sample size.

ENDOTHELIAL NUCLEOTIDE CATABOLISM

AND ADENOSINE PRODUCTION

Ryszard T. Smolenski, Z. Kochan, C. Page, R. McDouall, A-M.L. Seymour and M.H. Yacoub

Cardiothoracic Surgery
National Heart and Lung Institute
Harefield Hospital
Harefield
United Kingdom

Adenosine is a key metabolite in myocardial autoregulation and interactions between different types of cells in its production, metabolism and action are vital. In this study, adenosine and other purine catabolite production were analyzed in cultured umbilical vein endothelial cells under control conditions (normal O_2 and substrate supply) and with inhibitors of ATP formation (iodoacetate and oligomycin -1 +O). Inhibitors of adenosine deaminase - erythro-9 (2-hydroxy-3-nonyl) adenine (EHNA) and adenosine kinase 5´ iodotubercidin (ITu) were included where indicated. Nucleotides and catabolites were evaluated by HPLC in the cells and in the medium throughout 45 min of incubation at 37^0C. Total amounts or increments per culture flask (n=4-5, ± SEM) were as follows: In controls with EHNA, cellular ATP, ADP and AMP contents were 9.0 ± 0.1, 0.83 ± 0.14 and 0.15 ± 0.03 nmol, while in the medium increases in hypoxanthine, xanthine and uric acid content were 1.0 ± 0.2, 0.4 ± 0.1 nmol and 0.5 ± 0.2 nmol. Adenosine increased by $-/006 \pm 0.04$ nmol. With additional presence of ITu, adenosine rose by 0.9 ± 0.2 nmol ($p < 0.005$). After 45 min incubation with I+O and EHNA, ATP was totally depleted, cellular AMP rose to 3.0 ± 1.2 nmol and no IMP increase was found. Medium adenosine rose by 5.6 ± 0.2 nmol and the increase in sum of other purine catabolites was 2.8 ± 0.3 nmol. In conclusion, adenosine formed in the endothelial cells under normal O_2 and substrate supply is recycled via adenosine kinase. In ATP depleted cells, nucleotide catabolism proceeds predominantly via adenosine.

VASCULAR ADHESION MOLECULES IN HEART TRANSPLANTATION

Patricia M. Taylor, Marlene L. Rose and Magdi H. Yacoub

Department of Transplant Immunology
National Heart and Lung Institute
Harefield Hospital
Harefield, Middlesex
United Kingdom

Adhesion of leucocytes to vascular endothelium is a necessary step leading to the migration of cells into underlying tissues. Vascular adhesion molecules regulate this process and may play an important role in graft rejection. Immunocytochemical studies have been used to investigate the expression of vascular adhesion molecules (PECAM, ICAM, VCAM-1 and ELAM-1) as well as MHC class I, and DR antigen in acute myocardial rejection and transplant associated accelerated coronary sclerosis (TX-ACS).

Acute Myocardial Rejection

Control donor left atrium (n = 15) and endomyocardial biopsies from heart transplant patients with acute rejection (n = 15) were studied. Expression of antigens was determined on arteriole and venule endothelium, capillary endothelium, myocardial membrane and endocardium.

In donor heart, PECAM, ICAM-1, HHC class I and DR antigen were strongly expressed. VCAM-1 was variably expressed by endothelium from arterioles and venules but not capillaries and ELAM-1 was restricted to venules. During rejection, however, VCAM-1 and ELAM-1 were induced on capillary endothelium and ECAM-1, like MHC class I, was induced on the myocardial membrane. In addition expression of ICAM-1, VCAM-1, and DR antigen was increased on arterioles, venules and endocardium.

We suggest that the release of cytokines upregulates expression of these vascular adhesion molecules, enhancing adhesion and migration through endothelial cells into the tissue during episodes of acute rejection.

Immunogenicity of Coronary Arteries and the Development of Transplant Associated Accelerated Coronary Sclerosis

We have characterized the antigens present on endothelium from control coronary arteries from unused donor hearts (n = 4) or heart transplant recipients whose original disease did not involve the coronaries (n = 8), and from transplanted hearts taken at time of redo (n = 4) or at postmortem (n = 10).

All control coronary arteries strongly expressed PECAM, ICAM-1, ELAM-1 and MHC classI. Expression of VCAM-1 and DR antigen was more patchy. There was no significant difference in the expression of MHC or adhesion molecules between control arteries and those from transplant patients. Only 5/14 transplanted coronaries showed intimal cellular proliferation typical of TX-ACS. Two of these also showed evidence of intimitis, as did one other specimen.

Normal coronary arteries appear to be highly immunogenic prior to transplantation. This suggests the endothelium would need little, if any, stimuli to initiate as well as be a target for an immune response ultimately leading to TX-ACS. Factors such as MHC mismatch, a high frequency of precursor cytotoxic T cells and anti-endothelial antibodies (*Dunn et al., 1992*) would predispose the transplant recipient to the disease.

REFERENCES

Dunn, M.J., Crisp, S.J., Rose, M.L., Taylor, P.M., Yacoub, M.H., 1992, Anti-Endothelial Antibodies and Coronary Artery Disease After Cardiac Transplantation. *Lancet* 339:1566-1570.

FREE RADICAL PRODUCTS IN PLASMA DURING CORONARY SURGERY

Hannu J. Toivonen and *Markku Ahotupa

Department of Anesthesia
Helsinki University Central Hospital
Helsinki
Finland

*Department of Physiology
University of Turku
Turku
Finland

Ischemia-reperfusion is a strong inducer of free radical formation and lipid peroxidation in animal tissues. There is, however, some species differences and e.g. human tissues have very low free radical generating xanthine oxidase activity. Formation of free radical reaction products during human cardiac operations has been sporadically reported, but the issue has not been systematically studied.

We measured three different free radical reaction products, i.e. fluorescent chromolipids, diene conjugates and thiobarbituric acid (TBA) reactive material together with plasma ability to trap artificially (ABAP) generated free radicals (=antioxidant capacity) from nine patients during elective coronary artery bypass operations. Plasma samples were drawn at 13 time points covering the major interventions during the procedure, i.e. general anesthesia, heparin and protamine administration, extracorporeal circulation, cardioplegia administration and aortic crossclamping. The appearance of fluorescent chromolipids in urine was also studied. The patients were ventilated with 100% oxygen and drugs with antioxidant properties were avoided, but the home medication of beta-blockers, calcium antagonists and nitrates were given in the morning of the operation.

General anesthesia with fentanyl, pancuronium, diazepam and enflurane did not affect any of the measured parameters, neither did the administration of heparin at 300 IU/kg. When extracorporeal circulation was started, the plasma antioxidant capacity as well as the concentration of fluorescent chromolipids remained unchanged, but the concentration of diene conjugates stayed decreased throughout the perfusion period. After correction for hemodilution, fluorescent chromolipids were elevated after 5 minutes of extracorporeal circulation to 177 ± 14% (mean ± SEM), diene conjugates were elevated to 138 ± 12%, TBA derivatives to 152 ± 31% and plasma antioxidant capacity to 144 ± 12% of the awake value. Fluorescent chromolipid values remained at 156-177% of the awake value throughout the perfusion period and decreased to 130% ± 13 at one hour after the perfusion. Diene conjugates remained 123-144%, TBA derivatives at 86-147% and antioxidant capacity at 143-161% from the baseline during the perfusion period and these values decreased to 119 ± 5%, 103 ± 20, and 135 ± 9%, respectively, one hour after the perfusion. The TBA reactive material levels did not show statistically significant changes at any point. All the other parameters measured during the perfusion period were significantly elevated (ANOVA, followed by Newman-Keuls' test), whereas, none of these was significantly elevated, one hour after the termination of cardiopulmonary bypass. Hourly urine samples revealed that significant amounts of fluorescent chromolipids were immediately excreted in urine.

In conclusion, the amount of free radical products increases in plasma when extracorporeal circulation is introduced and stays elevated during the perfusion period. Fluorescent chromolipids appear to be more sensitive markers of free radical formation than diene conjugates or TBA reactive material. On the other hand, fluorescent chromolipids are excreted readily in urine, apparently decreasing the plasma values. Plasma antioxidant capacity increased parallel to the formation of free radical reaction products. General anesthesia or heparin administration did not affect plasma levels of the measured compounds, whereas, the most significant single challenger was the introduction of extracorporeal circulation (*Supported by the Academy of Finland, and the Juho Vainio Foundation, Finland*).

DOES ENDOTHELIUM PLAY AN ACTIVE ROLE ON THE MECHANISM OF CARDIOVASCULAR CALCIFICATION

Branko Tomazic

ADAHF, Paffenbarger Research Center
National Institute of Standards and Technology
Gaithersburg, MD 20899
U.S.A.

The title of the present contribution is actually a continuation of questions raised at the last ASI Vascular Endothelium Conference, 1990. The question: "What diseases are primarily diseases of endothelium?" remains as one not completely answered (*Brigham, 1991*). The endothelium is an important source of many biogenic factors such as cytokines that appear to be mediators of complex interactions (*Mantovani and Dejana, 1989*); some of these mediators may be natural promoters or inhibitors of a series of pathological processes. Another crucial function of endothelial cells is purine regulation (*Gordon, 1991*), which through ATP and ADP represent the main pool of phosphate. The mass transport through functional endothelium is a very important but not yet fully documented factor, particularly for inorganic species. Hammerson and Hammerson (1984) discussed the endothelium reaction pathways indicating that endothelium in its complexity controls reactions at membrane interface and mass transfer. The fact that lesion of cardiovascular tissues coincides with injuries of the endothelium raises an interesting possibility that the phenomenon of dystrophic calcification is directly or indirectly related to an end of function of endothelium. Hardening of arteries is commonly understood as deposition of cholesterol. An additional complication is pathologic calcification, and this phenomenon is often overlooked even though it takes place in many segments of cardiovascular system, including tissue-derived bioprostheses (*Schoen, 1989*). Pathologic calcification is a very complex process, which is certainly orchestrated by many biochemical factors including proteins, proteolipids, phospholipids and cholesterol. The mechanism of the biomineralization process is not certain, but it most likely involves the cell membrane elements and depends on the chemical state of extracellular vs. intracellular fluids. Their composition determine the chemical driving force for nucleation and crystallization of calcific deposits. The process is certainly mediated by a variety of known (pyrophosphate) and unknown natural inhibitors. The absence of natural inhibitors, which may be a result of cell injury, leads to pathological calcification. Comparative evaluation of the role of all these factors to elucidate the mechanism of cardiovascular calcification is not available. Yet, substantial information is provided by the multidisciplinal characterization of the inorganic component of cardiovascular deposits. The deproteinated inorganic component was characterized by chemical analyses, optical and structural techniques (SEM, EDX, polarizing microscopy, XRD,FTIR) and by solubility measurements. The composition of pathologic deposits approximate calcium-deficient, carbonate/phosphate substituted bioapatites (*Tomazic et al., 1988*). Thermodynamic solubility measurements convincingly eliminate hydroxyapatite (HAP, $Ca_5(PO_4)_3OH$, as the valid representative of cardiovascular deposits. The combined physicochemical information, supported by solubility measurements, strongly suggest octacalcium phosphate (OCP), $Ca_4H(PO_4)_3$, as a transient precursor in the formation of cardiovascular calcific deposits. The results of comparative *in vitro* and *in vivo* mineralization of bovine pericardium provided additional support that OCP may be a valid precursor in the formation of bioapatites, irrespective of the location of formation (*Tomazic et al., 1991*). The OCP precursor forms preferentially to HAP as a consequence of kinetic factors and the presence of inhibitors (*Eidelman et al., 1991*). The subtle difference in chemical composition of cardiovascular deposit elucidates the nature of the maturation process, which may be understood as time-dependent hydrolytic transformation of a precursor(s), the transformation rate being mediated by composition of ionic medium (*Tomazic et al., 1991*). The fact that the formation of calcific deposits in tissue-derived, endothelium-free bioprostheses is a relatively fast process, when compared with the similar process in living cardiovascular tissue, points to function of endothelium as an important growth factor to be critically evaluated. (*Supported in part by NHLBI grant HL30035*).

REFERENCES

Brigham, K.L., 1991, Vascular Endothelium: Physiological Basis of Clinical Problems. NATO ASI Series A 208:233-237.

Eidelman, M., Meyer, J.L. and Brown, W.E., 1991, Selective inhibition of crystal growth on octacalcium phosphate and hydroxyapatite by pyrophosphate at physiological concentrations. *J. Cryst. Growth* 113:643-652.

Gordon, J.L., 1991, Purine regulation of endothelial cells: relevance to pathophysiology. *Vascular Endothelium: Physiological Basis of Clinical Problems.* NATO ASI Series A 208:111-116.

Hammerson, F. and Hammerson, E., 1984, The ultrastructure of microvascular endothelial cell reactions to various stimuli. *Prog. Appl. Microcirc.* 6:91-108.

Mantovani, A. and Dejana, E., 1989, Cytokines as communicating signals between leukocytes and endothelial cells. *Immunol. Today* 10:370-375.

Schoen, F.J., 1989, Interventional and surgical Cardiovascular Pathology. *Clinical Correlations and Basic Principles.* W.B. Saunders Company, Philadelphia pp. 41-48.

Tomazic, B.B., Brown, W.E., Queral, L.A. and Sadovnik, M., 1989, Physicochemical characterization of cardiovascular calcified deposits. *Atherosclerosis* 69:5-19.

Tomazic, B.B., Mayer, I. and Brown, W.E., 1991, Ion incorporation into octacalcium phosphate hydrolyzates. *J. Cryst. Growth* 108:670-682.

Tomazic, B.B., Siew, C. and Brown, W.E., 1991, A comparative study of bovine pericardium mineralization: a basic and practical approach. *Cells and Materials* 1(3):231-241.

THE EFFECT OF INDOMETHACIN ON CYCLOSPORIN A- AND ITS SOLVENT-INDUCED INHIBITION OF ENDOTHELIUM-DEPENDENT RELAXATION AND THE DRUG-INDUCED CONTRACTION OF RABBIT ISOLATED ARTERIES

Meral Tuncer and Ersin Yaris

Department of Pharmacology
Faculty of Medicine
Hacettepe University
08100, Ankara
Turkey

This study was designed to investigate the effects of cyclosporin A preparation (Sandiummun) and its solvent (Cremophar-EL) on acetylcholine-induced endothelium-dependent relaxation of rabbit superior mesenteric artery and thoracic aorta, in vitro, in the absence and presence of indomethacin, a cycloxygenase inhibitor (10^5 M). The effect of indomethacin on the drug- and solvent-induced contractions of superior mesenteric artery was also evaluated.

Acetylcholine-induced endothelium-dependent relaxation was inhibited by cyclosporin A preparation (50 μg/ml) in both arteries but to a greater extent in the superior mesenteric artery. The solvent of cyclosporin A, in the concentration corresponding to that of the drug, caused less inhibition than cyclosporin A preparation on the relaxation of both arteries. Acetylcholine-induced relaxations were not affected by indomethacin. The inhibitory action of cyclosporin A preparation on the endothelium-dependent relaxation was significantly decreased by indomethacin only in the mesenteric artery. However, indomethacin had no effect on the solvent-induced inhibition of relaxation.

Cyclosporin A preparation (5 μg/ml) induced a slow increase in the tone of mesenteric artery. The solvent produced similar contractions of the artery. Removal of the endothelium did not change the contractile activity of the drug. The contractions induced by the drug and its solvent were significantly decreased by indomethacin.

The results suggest that Cremophor EL may contribute to the inhibitory action of cyclosporin A preparation on acetylcholine-induced endothelium-dependent relaxation and the smooth-muscle contracting effect of the drug in the superior mesenteric artery and that some contracting prostanoids may partly play a role in these actions of the drug and its solvent. * *This study was supported by Hacettepe University Research Grant no. HUAF 89 01 011 01.*

REFERENCES

Amorena, C., Castro, A., Muller, A., Villamil, M.P., 1990, *Clin. Sci*. 79:149-54.

Luscher, T.F., Vanhoutte, P.M., 1986, *Hypertension* 8:344-48.

Miller, V.M., Vanhoutte, P.M., 1986, *Am. J. Physiol*. 248:H432-H437.

Xue, H., Bukoski, R.D., McCarron, D.A., Bennette, W.M., 1986, *Transplantation* 43:715-8.

NITRIC OXIDE MEDIATED VASODILATION IN RESPONSE

TO INFLAMMATORY STIMULI

John B. Warren

Department of Applied Pharmacology
National Heart & Lung Institute
London SW3 6LY
United Kingdom

Nitric oxide, derived from many cell types, has been implicated in all three components of acute inflammation: vasodilation, edema formation, and leukocyte accumulation. Vasodilation, causing an initial increase in blood flow, seems intuitively a logical response to tissue injury. However, the precise importance of vasodilation to the subsequent events is not known, although it is established that vasodilation potentiates both edema formation and leukocyte accumulation in several examples of the inflammatory response.

The control mechanisms of vessel calibre of the microcirculation differ from those of the larger conduit arteries. In general, prostaglandin mediated stimulation of adenylate cyclase is more important in microvessels and nitric oxide stimulation of guanylate cyclase more important in large arteries. Although the skin is relatively unresponsive to agents such as nitroprusside or acetylcholine, the microcirculation does respond to nitric oxide derived from the inducible form of nitric oxide synthase. Both endotoxin and ultraviolet light B cause delayed onset erythema of the skin in animal models which can be prevented by inhibitors of the arginine-nitric oxide pathway or by pre-treatment with topical corticosteroids. Prostaglandins also contribute, as both endotoxin and ultraviolet light induced responses are attenuated by cyclooxygenase inhibitors. Arachidonic acid injected locally causes vasodilation in skin which is inhibited not only by indomethacin but surprisingly by nitric oxide synthase inhibitors. This suggests and interaction between the arachidonic acid pathway and the arginine-nitric oxide pathway.

Both the prostaglandins and the inducible form of nitric oxide synthase contribute to the vasodilation of the microcirculation that occurs in the models described above. The antagonism of these mechanisms contributes to the anti-inflammatory activity of aspirin-like drugs and corticosteroids.

REFERENCES

Warren, J.B., Coughlan, M.L., Williams, T.J., 1992 (in press). Endotoxin-induced hyperemia in rat skin involves nitric oxide and prostaglandin synthesis. *Br. J. Pharmacol.*

HYALURONAN RECEPTORS ON HUMAN ENDOTHELIAL CELLS - THE EFFECT OF CYTOKINES

D.C. West

Department of Immunology
University of Liverpool
Liverpool L69 3BX
England
United Kingdom

Hyaluronan is an extracellular glycosaminoglycan found in most, if not all, mammalian tissues. High levels of hyaluronan have been noted in many avascular tissues, transiently in tissue undergoing remodeling and morphogenesis, and at the periphery of tumors. Also, many pathological or inflammatory situations show local accumulation of hyaluronan eg. psoriasis, bleomycin induced lung damage and artherosclerosis. Hyaluronan is reported to regulate cell proliferation, adhesion, locomotion and differentiation, its effects being dependent on the size and concentration of hyaluronan and the cell type being studied. These effects are, in most cases, mediated by membrane by membrane receptors, such as CD44 or RHAMM.

In vivo studies have shown that high concentrations of high molecular weight hyaluronand ($>10^6$ Da) inhibit vessel formation in granulation tissue formation or cause regression of the pre-existing microvasculature. Low molecular weight hyaluronan oligosaccharides (2-10kDa) stimulate angiogenesis in the chick chorioallantoic membrane and in a rat wound healing model (West and Fan, unpublished data). *In vitro*, hyaluronan oligosaccharides, 3-16 disaccharides in length, specifically stimulate endothelial cell proliferation and migration. High molecular weight hyaluronan both inhibits endothelial cell proliferation and disrupts newly formed monolayers, suggesting that hyaluronan metabolism may play an important role in regulating angiogenesis. The present studies were performed to determine the native of human endothelial cell hyaluronan receptors, which appear to mediate the effects outline above and to examine the effect of cytokines on these receptors. CD44 is reported to be upregulated in activated T and B cells.

The present study has analyzed the binding of [^{3}H]-hyaluronan to the endothelial cell of both human (HUVEC, ECV304) and bovine (BA) origin. binding was in all cases and Scatchard analysis gave K_d values of 2.014×10^{-10}, 2.417×10^{-10} and 2.077×10^{-10}M for BA, ECV304 and HUVEC cultures, with 83,000, 152,000 and 160,000 receptors per cell. Neither prior hyaluronidase treatment nor digitonin permeabilization had any significant effect on receptor number. The three cytokines had varying effects on hyaluronan binding depending on cell- type. With BA cultures, all three cytokines significantly increased the level of hyaluronan binding (66- 93%). Although the increased level due to FGF is partly due to a 25% increase in cell number. Neither IL-1 nor TNF had any effect on binding by ECV304 cultures, but FGF increased binding by 46%. In this case, cell number was only increased by 115%. The HUVEC cultures showed a slight increase of binding (19%) with FGF treatment, but a 10% increase in cell number was also found. However, both IL-1 and TNF decreased binding by 22% and 30%, respectively.

SDS - PAGE analysis of [^{35}S]- methionine- labelled byaluronan- binding proteins, from control and treated cultures, showed the same general pattern of proteins for the three cell-types. Major bands were at 45 and 60 kDa, and appear to be increased by cytokine treatment in BA cultures. Fig 3. In ECV304 cells the 60 kDa band is decreased or absent after TNF and IL-11 treatment and increased by FGF. The 60 kDa band is the most prominent in HUVEC cells, and is down regulated in the presence of IL-1.

The binding of hyaluronan to BA cells was similar to that seen previously with cells in suspension, with a Kd of 2.014 x 10^{-10}M compared with 2x10^{-10}M found previously. However, in the previous study metabolically-labelled high molecular weight (10^6 Da) hyaluronan was used, which may account for the discrepancy in receptor numbers i.e. 83,000 compared to 2,000 found previously. Both the human endothelial cell- types had similar binding constants to BAs, but slightly higher receptor numbers. FGF increased receptor numbers in all three cell- types, although most notably in the BA cultures. This may be related to the chemotactic effect of this cytokine on endothelial cells. The effect of Il-1 and TNF varied with cell- type, being stimulatory in BAs and inhibitory in HUVEC cultures.

Methionine- labelled hyaluronan- binding proteins gave a similar electrophoretic pattern with all three cell- types, suggesting that the same proteins are involved in hyaluronan binding in both bovine and human endothelial cells. Two main proteins, 45 and 60kDa, are modulated by the cytokines. Previous studies have examined ^{125}I-labelled cell-surface hyaluronan "receptors". This analysis showed five major protein bands between 90 and 125 kDa, with less prominent proteins at 60 and 45 kDa. Thus, cytokine induced changes in these two proteins may be responsible for the modulation of surface hyaluronan binding. The more prominent surface "receptor proteins" can also be identified in the present gels, but no change is evident in these proteins.

Thus, cytokines can modulate endothelial hyaluronan- receptors. Their effect appears to be on the 60 and 45kDa "*receptors*" and not the high molecular weight "*receptors*" previously identified.

CYTOKINE UPREGULATION OF CD36, BUT NOT ICAM-1, INCREASES PLASMODIUM FALCIPARUM-INFECTED ERYTHROCYTE ADHERENCE TO MICROVASCULAR ENDOTHELIAL CELLS UNDER SHEAR CONDITIONS

Timothy M. Wick, J.K. Johnson, *R.A. Swerlick, **K.K. Grady and P. Mille

School of Chemical Engineering
Georgia Institute of Technology
Atlanta, GA 30332-0100
U.S.A.

*Department of Dermatology
Emory University School of Medicine
Atlanta, GA 30322
U.S.A.

**Malaria Branch
Centers for Disease Control
Atlanta, GA 30333
U.S.A.

Cytoadherence of parasitized red blood cells (PRBC) to microvascular endothelial cells (MEC) contributes to the pathogenesis of human cerebral malaria (*MacPherson et al., 1985*). We studied cytoadherence of two strains (HB3 and FC27) of *Plasmodium falciparum* PRBC to human MEC cultures under shear conditions to elucidate the pathways of adherence to MEC (*Wick and Louis, 1991*). Using anti-ICAM-1 and anti-CD36 antibodies, we found that HB3 PRBC bound exclusively to CD36 and ICAM-1 receptors. FC27 PRBC bound to MEC via CD36 (and at least one other pathway) but not to ICAM-1. Stimulation of MEC with interferon gamma (IFN-y), which selectively increases CD36 expression, elevated HB3 and FC27 PRBC adherence 64% and 71%, respectively. Conversely, MEC stimulation with tumor necrosis factor (TNF), which selectively upregulates ICAM-1, did not increase HB3 PRBC adherence, in contrast to previous data using large vessel endothelium. Downregulation of CD36 and ICAM-1 expression by phorbol 12, 113-dibutyrate completely abolished HB3 PRBC adherence. These data suggest that cytoadherence to microvascular EC is distinct from adherence to large vessel EC (*Berendt et al., 1989*). PRBC adherence to MEC is strain specific, regulatable by cytokines which increase CD36 expression, and abrogated by pharmacologic modulation of MEC receptors. These data have defined the relative contributions of both CD36 and ICAM-1 to PRBC binding to microvascular endothelium and have provided evidence for the presence of an additional adhesion mechanism. These novel results suggest to us that PBRC adherence in the microcirculation likely contributes to the occlusion and ischemic tissue damage associated with cerebral malaria via different mechanisms than cytoadherence to venous endothelium. Furthermore, these data suggest that in addition to antibody blocking of cell adhesion molecules, anti-cytokine (*Grau et al., 1989*) antibody therapy of pharmacological manipulation of endothelial cell receptor expression may reduce PBRC adherence to MEC and ameliorate the sequestration associated with human cerebral malaria.

REFERENCES

MacPherson et al., 1985, *Am. J. Pathol.* 119:385.
Wick and Louis, 1991, *Am. J. Trop. Med. Hyg.* 45:578.
Berendt, et al., 1989, *Nature* 341:57.
Grau, et al., 1989, *Int. Arch. Allergy Appl. Immunol.* 88:34.

THROMBOSPONDIN FROM ACTIVATED PLATELETS PROMOTES SICKLE ERYTHROCYTE ADHERENCE TO HUMAN MICROVASCULAR ENDOTHELIAL CELLS VIA CD36 AND INTEGRIN RECEPTORS

Timothy M. Wick, H.A. Brittain, *R.A. Swerlick
R.J. Howard and **J.R. Eckman

School of Chemical Engineering
Georgia Institute of Technology
Atlanta, GA 30332-0100
U.S.A.

*Department of Dermatology
Emory University School of Medicine
Atlanta, GA 30322
U.S.A.

**Department of Medicine
DNAX Research Institute
Palo Alto, CA 94304
U.S.A.

Adherence of erythrocytes to vascular endothelium likely contributes to the pathophysiology of episodic microvascular occlusion in patients with sickle cell disease (Wick, et al., 1987). In addition, coagulation activation has been reported in sickle patients during complications such as pain episodes (Francis and Johnson, 1991). Since platelets contain adhesion proteins (e.g. von Willebrand factor, thrombospondin and fibrinogen), we investigated whether factors released from activated platelets from sickle patients promote adherence of sickle erythrocytes to human microvascular endothelial cells (MEC) under flow conditions.

Activated sickle platelet supernatant (ASPS) promotes high levels of sickle erythrocyte adherence to MEC (55.4 ± 3.9 erythrocytes/mm^2) but only moderate adherence of normal erythrocytes to MEC (14.1 ± 0.7 erythrocytes/mm^2). The ASPS-mediated sickle erythrocyte adherence correlates with the platelet counts for 27 patients studied. When MEC are incubated with an antibody (OKM5) against CD36 (a thrombospondin receptor), platelet supernatant mediated sickle erythrocyte adherence is inhibited 86%, suggesting that thrombospondin is involved in adherence.

To further define the role of thrombospondin (TSP) in adherence, additional studies using purified TSP were performed. At a concentration of 0.2 μg/ml TSP in serum-free media, sickle erythrocyte adherence to MEC was 33.9 ± 2.7 erythrocytes/-mm^2, and was 6-fold greater than either sickle erythrocyte adherence in the absence of TSP or normal erythrocyte adherence in the presence of TSP. Doubling the concentration of TSP to 0.4 μg/ml proportionally increased adherence of sickle erythrocytes. TSP binds to CD36[3] and the vitronectin receptor integrin, $\alpha_v 3$[4]. Incubation of MEC with anti-CD36 antibody (OKM5) or anti-α_v monoclonal antibody inhibits TSP-mediated sickle erythrocyte adherence greater than 95%.

These data suggest that activated platelets release factors, including α-granule thrombospondin, which promote receptor mediated sickle erythrocyte adherence to microvascular endothelium. Such factors released during *in vivo* platelet activation could contribute to vaso-occlusive complications by promoting erythrocyte adherence and microvascular occlusion.

REFERENCES

Wick, et al., 1987, *J. Clin. Invest.* 80:905.
Francis and Johnson, 1991, *Blood* 77:1405.
Asch, et al., 1987, *J. Clin. Invest.* 79:1054.
Lawler and Hynes, 1989, *Blood* 74:2022.

POST-PRANDIAL CHYLOMICRON CREAMING, C-REACTIVE PROTEIN AND ENHANCED COMPLEMENT ACTIVATION

Nicholas Wickham, J. Cormak, G.M. Vercellotti
D.E. Hammerschmidt and H.S. Jacob

Department of Hematology
St. George's Hospital
London
United Kingdom

Department of Medicine
University of Minnesota Hospital
Minneapolis, Minnesota
U.S.A.

The acute-phase reactant, c-reactive protein (CRP): 1) activates human complement (*Volanakis, J.E., 1982*); 2) causes creaming *in vitro* of lipid micro-emulsions (*Rowe, I.R., Soutar, AK, and Pepys MB*, 1986); 3) is thought to underlie fat micro-embolism *in vivo* in sick patients receiving parenteral nutrition containing lipid (*Hulman, G. et al., 1982*) as a chronic high saturated fat intake is a risk factor for ischemic heart disease, acute myocardial infarction (MI) frequently follows large meals, and complement activation has been shown to take place during MI (*Langlois, P.F., and Gawryl, M.S., 1992*), we wondered whether chylomicrons (CM) might potentiate CRP-induced complement activation.

CM were separated from whole edta-blood taken from health volunteers after a fatty meal, light-scattering measurements, at 700 nm, were used to achieve a final experimental CM concentration equivalent to that of the initial lipemic plasma, human CRP (*Dr. H. Gewurz, Chicago*) at 100 MG/L (NR < 5) caused instant (calcium-dependent) creaming of CM both in plasma and after separation. CM incubated with normal human plasma had no effect on C3a generation compared to control, while CRP alone increased C3a generation by 33% (n=8, p=0.01). This value, however, more than doubled to 85% in the presence of CM (P < 0.05). Interestingly, CM did not enhance the effect of heat-modified CRP, although it also caused creaming and activated complement to a similar extent to unmodified CRP. Pilot experiments using rabbit neutropenia as a sensitive model for *in vivo* complement activation, lend some support to the above data.

Complement activation by C-reactive protein, but not necessarily creaming, and its enhancement by CM were both blocked by monoclonal antibodies against a variety of CRP epitopes (*Ying S-C et al., 1989*). Although there was a trend suggesting that the presence of CM with CRP generated higher values of C3a than CRP alone.

CM may thus constitute a dual risk; first, by creaming and potentially blocking the micro-vasculature, and secondly, by enhancing CRP-induced complement activation, they may initiate and aggravate acute inflammatory and ischemic events.

These findings may: 1) explain partly the link between thrombosis (both myocardial (*Spodick, D.H., 1986*; *Syrjanen, J., 1990*) and cerebral (*Syrjanen, J., 1990*;

Syrjanen, J. et al., 1986) with preceding infection, and the fact that MI is more extensive in both animals and humans who have lipemic plasma (*Osborne, J.A., 1987*) provide an insight into the circadian variation in thrombosis (*Kolata, G., 1986; Muller, J.E., 1986; Mulcahy, D., Purcell, H. and Fox, K.*) in that the morning peak incidence in infarction and strokes would coincide with the breakfast post-prandial CM peak.

REFERENCES

Hulman, G. et al., 1982, Agglutination of intralipid by sera of acutely ill patients. *Lancet* Dec 25:1426-1427.

Kolata, G., 1986, Research News: Heart Attacks at 9 a.m.. *Science* 233:417-8.

Langlois, P.F., Gawryl, M.S., 1988, Detection of the terminal complement complex in patient plasma following acute myocardial infarction. *Atherosclerosis* 70(1-2):95-105.

Mulcahy, D., Purcell, H., and Fox, K. Should we get up in the morning? Observations on circadian variations in cardiac events. *Br Heart J.* 65:299-301

Muller J.E., 1985, Circadian variation in onset of acute myocardial infarction. *New Eng. J. Med.* 313-:1315-22.

Osborne, J.A. et al., 1986, Increased severity of acute myocardial ischaemia in experimental athero--sclerosis. *Ht Vessels* 3(2):73-8.

Rowe, I.R., 1986, Agglutination of intravenous lipid emulsion (*Intralipid*) and plasma lipoproteins by C-reactive protein. *Clin. Exp. Immunol.* 66(1):241-7.

Spodick, D.H., 1986, Infection and infarction - acute viral (and other) infection in the onset, patho-genesis, and mimicry of acute myocardial infarction. *Am. J. Med.* 81(4):661-8.

Syrjanen J., 1990, Vascular Diseases and Oral Infections. *J. Clin. Periodontal* 17(7):497-500.

Syrjanen J. et al., 1986, Association between cerebral infarction and increased serum bacterial anti-body levels in young adults. *Acta Neurol Scand* 73(3):273-8.

Volanakis, J.E., 1982, Complement activation by C-reaction protein complexes. *Ann NY Acad. Sci.* 389:235-249.

Ying, S-C et al., 1989, Identification and partial characterization of multiple native and neoantigenic epitopes of human C-reactive protein by using monoclonal antibodies. *J. Immunol.* 143(1):221-228.

VIII. PARTICIPANTS

PARTICIPANTS

Andreasen, Frederic, M.D.
Department of Pharmacology
University of Aarhus
The Bartholin Building
DK 8000, Aarhus C
DENMARK

Armaganidis, A., M.D.
Critical Care Medicine
Evangelismos General Hospital
45-47 Ipsilandou Street
GR 106, Athens
GREECE

Bassenge, Eberhard, M.D.
Angewandte Physiologie
Albert Ludwig University
Hermann Herder str. 7
7800 Freiberg
GERMANY

Bastaki, Maria
Department of Pharmacology
University of Patras
Medical School
26 500 Rio
Patras, GREECE

Baydanoff, Stephan, M.D.
Head of Biology
Kiril and Methodi str
Medical University
Pleven 5800
BULGARIA

Behrakis, P.K., M.D.
Intensive Care Unit
Red Cross Hospital-Athens
Erithros Stavros No. 1
11526 Athens
GREECE

Bizros, Rena, Ph.D.
Biochemical Engineering
Rensselaer Polytechnic
Troy, N.Y., USA

Bogle, Richard
Biomedical Sciences
King's College London
Campden Hill Road
Kensington/London W8 7AH
UNITED KINGDOM

Burke-Gaffney, Anne
Children's Research Center
Our Lady's Hospital for
 Sick Children
Dublin 12, IRELAND

Callow, Allan D., M.D., Ph.D.
Department of Surgery
Washington University Medical Ctr.
4960 Audubon Avenue
St. Louis, MO, USA

Catravas, John D., Ph.D.
Dept. of Pharmacolaogy and
 Toxicology
The Medical College of Georgia
Agusta, GA 30912-2300, USA

Cecchelli, Romeo, Ph.D.
Service of Research
Institute of Pasteur, 20
bd Louis XIV, Lille
FRANCE

Chatzicondi, Olga, Ph.D.
Dept. of Pharmacology
University of Patras
 Medical School
26500 Patras, GREECE

Chen, Xilin, M.D.
Department of Molecular
 Physiology and Biophysics
Baylor College Medicine
One Baylor College
Houston, TX 77030, USA

Chester, Adrian, M.D.
Department of Thoracic
 and Cardiac Surgery
Harefield Hospital
Harefield, Middlesex UB9 6JH
UNITED KINGDOM

Colak, Rukiye
Hacettepe University
P.I. 137, Maltepe
Ankara, TURKEY

Cooke, Brian M.
Department of Hematology
University of Birmingham
Edgbaston, Birmingham
UNITED KINGDOM

Cunha-Ribeiro, Luis M., M.D.
Department of Physiology
Hospital San Joao
4200 Porto
PORTUGAL

De Bault, Lawrence E., Ph.D.
Department of Pathology
University of Oklahoma
Health Sciences Center
Oklahoma City, OK
USA

De Kimpe, Sjef J.
Faculty of Pharmacy
University of Utrecht
Sorbonnelaan 16, 3584 CA
Utrecht
THE NETHERLANDS

Dehouck, M.P., Ph.D.
Service of Research
Institute Pasteur
Institute Pasteur
20 bd Louis XIV
Lille
FRANCE

Demetriou, Achilles, M.D.,
 Ph.D.
VA Medical Center
1310 24th Avenue South
Nashville, TN 37212-2637
USA

Demir, Ramazan, M.D.
Department of Histology
 and Embryology
Akdeniz University
07070 Arapsuyu, Antalya
TURKEY

Eksioglu, Umit, M.D.
A. Ayranci 06540
Ankara
TURKEY

Engel, Leslie, Ph.D.
Health Sciences Dept.
Monsanto Company
800 N. Lindbergh
 Boulevard
St. Louis, MO 63167
USA

Fehr, Jorg, M.D., Ph.D.
Division of Hematology
Medicine, A Hof 149
University Hospital
CH-8091 Zurich
SWITZERLAND

Fellstrom, Bengt, M.D., Ph.D.
Department of Internal
 Medicine, Renal Unit
University Hospital
S75185 Uppsala
SWEDEN

Flitney, Frederick W., Ph.D.
Department of Biology and
 Preclinical Medicine
University of St. Andrews
Bute Medical Building, Fife
KY169TS
UNITED KINGDOM

Frangos, John, Ph.D.
Department of Chemical
 Engineering
PENNSTATE
133 Fenske Laboratory
University Park, PA
USA

Freeman, Bruce A., Ph.D.
Dept. of Anesthesiology
University of Alabama
 at Birmingham
941 Tinsley Hamson Tower
Birmingham, AL 35294
USA

Garlanda, Cecilia, M.D.
Department of Research
 and Biology
Mario Negri, Via Eritrea 62
Milan
ITALY

Giavazzi, Raffaella, Ph.D.
Laboratory Negri Bergamo
Via Gavazzeni 11
24100 Bergamo
ITALY

Gordon, John, Ph.D.
Director of Research
British Bio-Technology Limited
Watlington Road, Cowley
Oxford, OX4 5LY
UNITED KINGDOM

Gorski, Andrew, Ph.D.
Department of Immunology
Transplantation Institute
Warsaw Medical School 02-006
Warsaw 22
POLAND

Gurtner, Gail H., M.D.
Division of Pulmonary
 and Critical Care Medicine
New York Medical College
Valhalla, NY, USA

Senaldi, Giorgio, M.D.,
 Ph.D.
Department of Pathology
University of Geneva
1, rue Michel-Servet
CH-1211, Geneva 4
SWITZERLAND

Haralabopoulos, George
Dept. of Pharmacology
University of Patras
 Medical School
26 500 Rio, Patras
GREECE

Hardy, Medora M., Ph.D.
Health Sciences Dept.
Monsanto Company
800 N. Lindbergh
 Boulevard
St. Louis, MO 63167
USA

Harlan, John M., M.D.
Harborview Medical Center
University of Washington
325 Ninth Street
Seattle, WA 98104
USA

Hornebeck, W., Ph.D.
Laboratory of Biology
Faculty of Medicine
University of Paris XII
8 rue du General Sarrail
94010, Creteil Cedix
FRANCE

Ibarra-Illarina, Corazon, M.D.
Triad Medical Center
4600 Kietzke Lane Suite M-242
Reno, Nevada 89502
USA

Jacobs, M., M.D.
Dept. of Pharmacology
Royal Free Hospital School
 of Medicine
Rowland Hill Street
London SW3 2PF
UNITED KINGDOM

Kefalides, Nicholas, A., M.D.,
 Ph.D.
Chief, Connective Tissue
 Research Section
University of Pennsylvania
 School of Medicine
University City Science Center
3624 Market Street
Philadelphia, PA 19104
USA

Kuhn, Michaela, M.D.
Dept. of Pharmacology
Institute of
 Peptide-Forschung
Feodor-Lynen street
313000 Hannover 61
GERMANY

Langleben, David, M.D.
Jewish General Hospital
3755 Ste-Catherine
Montreal, Quebec H3T 1E2
CANADA

Luscher, Thomas F., M.D.
Division of Cardiology
University Hospital
CH-3010 Bern
SWITZERLAND

Macarthur, Heather
The William Harvey Research
Institute
St. Bartholomew's Hospital
 Medical College
London
UNITED KINGDOM

Madge, Lisa
Department of
 Biomedical Sciences
King's College London,
 Kensington
London, W8 7AH
UNITED KINGDOM

Mantovani, Alberto, M.D.
Institute of Mario Negri
Via Eretria 62
20157 Milan
ITALY

Maragoudakis, Michael, Ph.D.
Dept. of Pharmacology
University of Patras
Patras
GREECE

Marczin, Nandor, M.D.
Dept. of Pharmacology
 and Toxicology
The Medical College
 of Georgia
Augusta, GA 30912-2300
USA

McMurdo, Lorraine
The William Harvey
 Research Institute
St. Bartholomew's
 Hospital Medical College
Charterhouse Square
London EC1M 6BQ
UNITED KINGDOM

Moser, Rene, M.D.
Division of Hematology
Department of Medicine
A Hof 143, University
 Hospital Zurich
CH8091 Zurich
SWITZERLAND

Murad, Ferid, M.D., Ph.D.
VA Medical Center
380 Miranda Avenue
Palo Alto, CA 94304
USA

Neisler, Hugh M., Ph.D.
Naval Aerospace Medical
 Research Laboratory
Pensacola, FL 32508
USA

O'Neil, Greg, Ph.D.
Thoracic and Cardiac
 Surgical Unit
Harefield Hospital
Harefield, Middlesex UB9 6JH
UNITED KINGDOM

Orfanos, Stylianos, E., M.D.,
 Ph.D.
Royal Victoria Hospital
Critical Care Division
687 Pine Avenue, West
Montreal
CANADA H3A1A1

Palmer, Richard, Ph.D.
The Wellcome Research
 Laboratory
Langley Court
Beckenham Kent, BR3 3BS
UNITED KINGDOM

Papaioannou, Stamatis E., Ph.D.
Department of Pharmacy
University of Patras
School of Health Sciences
Rio, Patras 261-10
GREECE

Papapetropoulos, Andreas
Dept. of Pharmacology
 and Toxicology
The Medical College
 of Georgia
Augusta, GA 30912-2300
USA

Pesmen, Arsu
Hacettepe University
P.K. 137 Maltepe
Ankara
TURKEY

Rao, Prakash, N., Ph.D.
Director, Transplant Research
Division of Liver
 Transplantation
Department of Surgery
Mt. Sinai Medical Center
Box 1104
19 East 98th Street
New York, NY 10029-6574

Rose, Marlene, Ph.D.
Department of Thoracic
 and Cardiac Surgical Unit
Harefield Hospital
Harefield, Middlesex UB9 6JH
UNITED KINGDOM

Roussos, Charis, M.D.
Department of Critical
 Care Medicine
Evangelismos General Hospital
Athens
GREECE

Ryan, Una S., Ph.D.
Director of Health Sciences
Monsanto Company
800 N. Lindbergh
 Boulevard
St. Louis, MO 63167
USA

Sakkoula, Eleni
Dept. of Pharmacology
University of Patras
 Medical School
26 500 Rio
Patras
GREECE

Scott, T.M., Ph.D.
Memorial University
 of Newfoundland
Division of Basic Medical
 Sciences
The Health Sciences Center
St. John's Newfoundland,
 A1B 3V6
CANADA

Sgoutas, Demetrious, S., Ph.D.
Department of Pathology
Emory University School
 of Medicine
Atlanta, GA 30322
USA

Soydan, Inan, M.D.
Department of Pathology
Ege University
Bornova, Izmir
TURKEY

Soydan, Saliha, M.D.
Department of Pathology
Ege University
Bornova, Izmir
TURKEY

Taylor, Patricia M., Ph.D.
Dept. of Immunology
Thoracic and Cardiac
 Surgery
Harefield Hospital
Harefield UB9 6JH
UNITED KINGDOM

Toivonen, Hannu, M.D., Ph.D.
Dept. of Anesthesiology
Helsinki University
 Hospital
Helsinki
FINLAND

Tomazic, Branko B., Ph.D.
American Dental
 Assoc. Health Foundation
Paffenbarger Research Center
National Institute of
 Standards and Technology
Gaithersburg, MD 20899
USA

Tsopanoglou, Nickolaos S.
Dept. of Pharmacology
University of Patras
 Medical School
26 500 Rio
Patras
GREECE

Tu, Anthony, T., Ph.D.
The Department of
 Biochemistry
Colorado State University
Fort Collins, CO 80523
USA

Tuncer, Meral, M.D., Ph.D.
The Department of
 Pharmacology
Faculty of Medicine
Hacettepe University
06100 Ankara
TURKEY

Tzanela, Marinella, M.D., Ph.D.
2021 Atwater Street
Apt. 1204
Montreal PQ
CANADA

Voyno-Yasenetskaya,
 Tatyna, Ph.D.
Cardiovascular Research
 Institute
School of Medicine
University of California
San Francisco, CA
USA

Warren, John, M.D.
Department of Applied
 Pharmacology
National Heart & Lung
 Institute
Dovehouse Street, London SW3
6LY
England
UNITED KINGDOM

West, David C., Ph.D.
Department of Immunology
University of Liverpool
Box 147, Liverpool L69 3BX
UNITED KINGDOM

Wick, Timothy, Ph.D.
School of Chemical
 Engineering
Georgia Institute of
 Technology
Atlanta, GA 30332-0100
USA

Wickham, Nicholas, Ph.D.
Department of Cellular and
 Molecular Science
St. George's Hospital
 Medical School
London SW17 ORE
UNITED KINGDOM

NATO ASI
VASCULAR ENDOTHELIUM
PHYSIOLOGICAL BASIS OF CLINICAL
PROBLEMS II
JUNE 20-30, 1992
PARADISE HOTEL, RHODES, GREECE

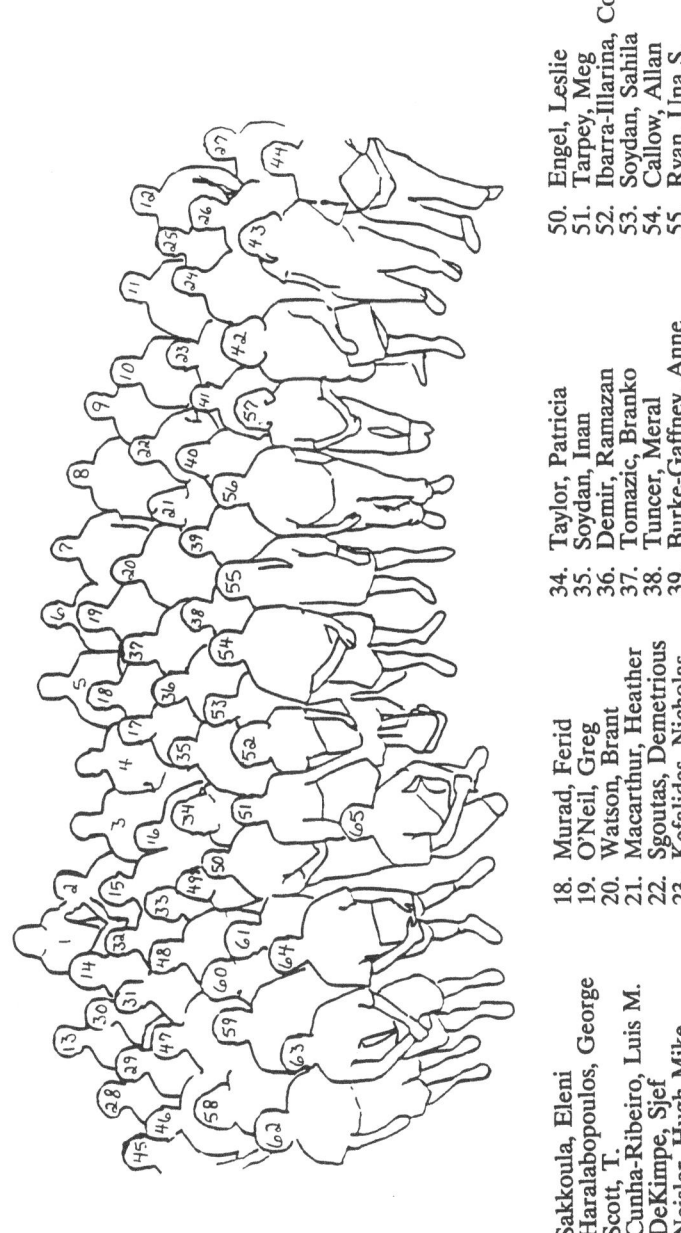

1. Sakkoula, Eleni
2. Haralabopoulos, George
3. Scott, T.
4. Cunha-Ribeiro, Luis M.
5. DeKimpe, Sjef
6. Neisler, Hugh Mike
7. Freeman, Bruce
8. Baydanoff, Stephan
9. Demetriou, Achilles
10. Rao, Prakash
11. Andreasen, Frederick
12. Toivonen, Hannu
13. Orfanos, Stylianos
14. Tsopanoglou, Nikolaos
15. Papapetropoulos, Andreas
16. Armaganidis, Apostolos
17. Eskioglu, Umit

18. Murad, Ferid
19. O'Neil, Greg
20. Watson, Brant
21. Macarthur, Heather
22. Sgoutas, Demetrious
23. Kefalides, Nicholas
24. Cooke, Brian
25. Marczin, Nandor
26. Chen, Xilin
27. Senaldi, Georgio
28. Bizios, Rena
29. Hornebeck, W.
30. Gorski, Andrew
31. Cecchelli, Romeo
32. Langleben, David
33. Tzanela, Marinella

34. Taylor, Patricia
35. Soydan, Inan
36. Demir, Ramazan
37. Tomazic, Branko
38. Tuncer, Meral
39. Burke-Gaffney, Anne
40. McMurdo, Lorraine
41. Jacobs, Michael
42. Skaparni-Dimopoulou, Efi
43. Kuhn, Michaela
44. Tu, Anthony
45. West, David
46. Wickham, Nicholas
47. Bogle, Richard
48. Warren, John
49. Dehouck, M.

50. Engel, Leslie
51. Tarpey, Meg
52. Ibarra-Illarina, Corazon
53. Soydan, Sahila
54. Callow, Allan
55. Ryan, Una S.
56. Catravas, John
57. Giavazzi, Raffaella
58. Madge, Lisa
59. Bassenge, Eberhard
60. Chatzicondi, Olgi
61. Roupp, Mary Ann
62. Hardy, Medora
63. Harlan, John
64. Flitney, Frederick
65. Fellstrom, Bengt

IX. INDEX

INDEX

The manufacturer's authorised representative in the EU is Springer
Nature Customer Service Centre GmbH, Europaplatz 3, 69115 Heidelberg,
Germany. If you have any concerns regarding our products, please
contact ProductSafety@springernature.com

Printed and bound by CPI Group (UK) Ltd, Croydon, CR0 4YY

29/04/2026
02099460-0019